INTRODUCTION TO
TELEPHONE SWITCHING

INTRODUCTION TO TELEPHONE SWITCHING

Bruce E. Briley
Bell Telephone Laboratories
Naperville, Illinois

1983

ADDISON-WESLEY PUBLISHING COMPANY, INC.

Advanced Book Program
Reading, Massachusetts

London • Amsterdam • Don Mills, Ontario • Sydney • Tokyo

Library of Congress Cataloging in Publication Data

Briley, Bruce Edwin, 1936-
   Introduction to telephone switching.

   Includes bibliographies and index.
   1. Telephone systems. 2. Telephone, Automatic.
I. Title.
TK6397.B67   1983   621.385'7   83-8835
ISBN 0-201-11246-9

.BCDEFGHIJ-MA-89876

TO MY WIFE, MARILYN

CONTENTS

ILLUSTRATION LIST   xiii

PREFACE   xix

1. BASIC TELEPHONY IN BRIEF   1
  1.1 Introduction, 1
  1.2 The Telephone Station, 5
  1.3 The Switching Interface, 7
  1.4 Transmission, 9
       Subscriber Loop, 9
       Interoffice Facilities, 9
  1.5 The Switching Machine, 10

  Exercises, 10

  Reading List, 11

2. WHY SWITCH?   13
  2.1 Preliminaries, 13
  2.2 Why Switch Centrally?, 14
  2.3 The Anatomy of a Simple Call, 19
  2.4 Switching Principles, 22
  2.5 Real-Time Interactive Operation, 25
  2.6 Switching Practices, 26

  Exercises, 29

  Reading List, 30

3. CURRENT SYSTEMS   31
  3.1 Local and Tandem, 31
       Step by Step (SXS), 32
       Panel, 37
       Crossbar, 41
           No.1 Crossbar, 42
           No.5 Crossbar, 47
           No.5A Crossbar, 49
       ESS, 49

No. 1 ESS, 50
No. 1A ESS, 63
No. 2 ESS, 65
No. 2A ESS, 66
No. 3 ESS, 67
No. 2B ESS, 68
No. 5 ESS, 68
3.2 Toll, 68
No. 4 ESS, 71
No. 5 ESS, 76
3.3 Operator, 77
Manual Boards, 79
Automatic Call Distributors, 82
TSPS, 82
Automatic Intercept System, 84

Exercises, 85

Reading List, 86

4. CURRENT FEATURES AND APPLICATIONS   89

4.1 Maintenance, 89
4.2 The Hierarchy, 91
4.3 Charging, 95
4.4 Signaling, 100
Station-Office, 102
Interoffice, 102
Conventional Techniques, 102
Common Channel Interoffice Signaling (CCIS), 106
4.5 Numbering Plan, 109
Domestic, 109
International DDD, 111
4.6 Expanded 911, 114
4.7 Advanced Mobile Phone Service (AMPS), 115

Exercises, 116

Reading List, 117

5. NETWORKS 119

5.1 Generalities, 119
5.2 Principles, 119
5.3 Nonblocking Networks, 123
5.4 Blocking Networks, 127
5.5 Simulation Techniques, 132
5.6 Network Technology, 133
    Electromechanical, 133
      Open Contact, 133
      Closed Contact, 134
    Electronic, 134
      Space Division, 135
      Time Division, 136
5.7 Techniques, 136
    Space Division, 136
    Time Division, 137
5.8 Trunking, 139
    Traffic Definitions and Characteristics, 139
    Traffic Principles, 148
      Lost Calls Held (LCH), 153
      Lost Calls Cleared (LCC), 154
      Lost Calls Delayed (LCD), 157
    Finite Sources, 165
    Limited Availability, 165
    Nonrandom Traffic, 165
    Delay Engineering, 168
    Extreme Value Engineering, 168
    Importance of Size, 168
5.9 Path Hunt, 168
5.10 Rearrangeable Networks, 170
5.11 Analysis of a Practical Network, 171
5.12 Nonseries-Parallel Networks, 174
5.13 Broadcast Networks, 176
5.14 Analysis of Time-Space Networks, 177

Exercises, 177

Reading List, 180

6. STORED-PROGRAM CONTROL   183

6.1 History, 183
6.2 Characteristics of Stored-Program Telephone Processors, 185
6.3 Why Assembly Language?, 187
6.4 Architectures, 189
6.5 Other Possible Stored-Program Machine Architectures, 191
        Multiprocessors, 192
        Language-Oriented Machines, 193
        Pipe-Lining and Cache Stores, 194
6.6 An Example of No. 1 ESS Code, 195

Exercises, 198

Reading List, 199

7. WORLD SYSTEMS   201

7.1 Generalities, 201
7.2 Details, 201
        United States, 202
            General Telephone and Electronics, 202
            North Electric, 203
            Stromberg-Carlson, 203
            Vidar, 203
        Canada, 204
            Northern Telecom, 204
        United Kingdom, 204
        France, 205
        West Germany, 205
        Sweden, 206
        The Netherlands, 206
        Italy, 206
        Japan, 207
        International Telephone and Telegraph, 207
7.3 Futures, 208

Exercises, 209

References, 209

8. ECONOMICS UNDER REGULATION   215

8.1 Introduction, 215
8.2 Demand Analysis, 215
      Competition, 218
      Price Umbrellas, 218
      Tariffs, 218
      Demand Estimation, 218
8.3 Engineering Economy, 220
      The Time Value of Money, 222

Exercises, 228

Reading List, 228

GLOSSARY   229

INDEX   249

ILLUSTRATION LIST

1-1 The Station Set, 7
1-2 The Line Interface, 8

2-1 The Need for Centralized Switching, 14
2-2 Non-Centralized Switching, 15
2-3 Centralized Switching: Non-blocking, Redundant, 16
2-4 Centralized Switching: Non-blocking, 16
2-5 Centralized Switching: Blocking, 17
2-6 Simplified Automatic Switching Machine, 18
2-7 Anatomy of a Simple Intraoffice Call, 20
2-8 Cross-Section of a Small Office, 27
2-9 Cumulative Distribution of Local Exchange Buildings and Lines, 28
2-10 Distribution of Machine-Type Usage, 29

3-1 The Tandem Office, 32
3-2 SXS Terminal Bank, 34
3-3 SXS Switches, 34
3-4 SXS Switch Bay, 35
3-5 A Call in a 10k SXS Office, 35
3-6 Panel Bay, 38
3-7 Panel Organization, 39
3-8 Panel Office Detail, 39
3-9 A 10X20 Crossbar Switch, 43
3-10 The Crossbar Switch Mechanism, 44
3-11 A Crossbar Switch Bay, 45
3-12 No. 1 Crossbar Organization, 46
3-13 No. 5 Crossbar Organization, 48
3-14 No. 1 ESS Genesis, 50
3-15 No. 1 ESS Organization, 52
3-16 The Sealed Reed Switch, 52
3-17 Exploded View of Ferreed, 53
3-18 Flux Paths, 53
3-19 An 8X8 Ferreed Switch, 54
3-20 No. 1 ESS Network, 55
3-21 No. 1 ESS Line-Link Network, 55
3-22 No. 1 ESS Trunk-Link Network, 56
3-23 The Ferrod Sensor, 56
3-24 No. 1 ESS Block Diagram, 57
3-25 The Twistor Element, 58

3-26 The Twistor Memory, 58
3-27 The Twistor Card, 59
3-28 The Twistor Store, 60
3-29 The Ferrite Sheet, 61
3-30 No. 1 ESS Intraoffice Call, 61
3-31 No. 2 ESS Genesis, 66
3-32 No. 2 ESS Network, 67
3-33 Toll Machine Genesis, 69
3-34 Typical Trunk Patterns, 70
3-35 Large Toll Office Advantages, 71
3-36 Time-Division Switch Elements, 72
3-37 A Simple PCM Network, 73
3-38 A Large PCM Network, 74
3-39 No. 4 ESS Functional View, 75
3-40 Simplified TST Path Set Up, 76
3-41 No. 5 ESS System, 78
3-42 Operator Census History, 78
3-43 Early Central Office, 79
3-44 The "Law Board," 80
     (The above 2 photos reprinted with permission from HARPER & ROW,
     Beginnings of Telephony, by F. L. Rhodes, 1929,
     © HARPER & BROS. PUBLISHERS, 1929; renewed 1957 by Eleanor
     A. Rhodes. Permission also granted by AT&T.)
3-45 An Early Rural Central Office, 80
3-46 A "Coed" Central Office, 81
3-47 A Large PBX, 81
3-48 The Traffic Service Position System, 83
3-49 TSPS Consoles, 84

4-1 Switching Plan (Basic Principle), 92
4-2 Distance Dialing Network, 92
4-3 Switching Plan (Routing Pattern), 93
4-4 Routing Example, 94
4-5 Zoning, 96
4-6 Message Registers, 96
4-7 Paper Tape Recorder, 97
4-8 Free Directory-Assistance Operator Requirements, 98
4-9 Directory-Assistance Users, 99
4-10 Signaling, 100

4-11 Signaling for a Typical Connection, 101
4-12 Touch-Tone Frequencies, 102
4-13 Station-Office Signaling, 103
4-14 Interoffice Signaling, 104
4-15 Multifrequency Signals, 105
4-16 Relative Signaling Speeds, 105
4-17 Interoffice Supervision, 106
4-18 Common Channel Interoffice Signaling, 107
4-19 Non-Associated CCIS, 108
4-20 Numbering Plan Areas, 110
4-21 Syntax of Distance Dialing, 111
4-22 Bell System Overseas Service, 112
4-23 World Numbering Zones, 113
4-24 Advanced Mobile Phone Service, 115
4-25 AMPS Channel Reuse, 116

5-1 Nonblocking, Full Access (Strict), 120
5-2 Blocking Full Access, 121
5-3 Origin of Network Terms, 122
5-4 Single-Stage Nonblocking Network, 124
5-5 Three-Stage Space-Division Network, 125
5-6 Clos Reasoning, 125
5-7 Nonblocking Network Crosspoint Count, 127
5-8 Three-Stage Clos Network, 128
5-9 Lee's Method, 129
5-10 Three-Stage Network Example, 129
5-11 Blocking: Eight-Customer Example, 130
5-12 8X8 Network Example, 131
5-13 Remreed Crosspoints, 135
5-14 PNPN Characteristics, 136
5-15 Time-Division Multiplexing, 137
5-16 Modulation Scripts, 138
5-17 Meaning of Intensity in Erlangs, 140
5-18 Example of Calling-Rate Variation, 142
5-19 Holding-Time Variation (Local Calls), 142
5-20 Example of Traffic-Intensity Variation, 143
5-21 Example of Call Variation, 143
5-22 CO Load-Time Relationship, 144
5-23 Seasonal Load Variation, 144

5-24 Busy-Hour Load Variation, 145
5-25 Seasonal Variation (Local Calls), 145
5-26 Seasonal Variation (Toll Calls), 146
5-27 Busy-Hour Load Distribution, 146
        (The above 10 figures reprinted with permission from AT&T,
        Switching Systems, 1961, © AT&T, 1961.)
5-28 Traffic H-Diagram, 148
5-29 LCH Curves, 155
5-30 LCH Curves (Contracted Scale), 156
5-31 LCC Curves, 158
5-32 LCC Curves (Contracted Scale), 159
5-33 LCD Curves, 162
5-34 LCD Curves (Contracted Scale), 163
5-35 Blocking Formula Comparison, 164
5-36 Effect of Variance-to-Mean Ratio, 166
        (Reprinted with permission from *Telephony*, August 2, 1971,
        "The Mathematics of Telephone Traffic Distribution," by
        R. R. Mina, p. 26, © 1971 by TELEPHONY PUBLISHING CORP.,
        55 E. Jackson Blvd., Chicago, IL 60604.)
5-37 Servers Versus Occupancy, 169
5-38 Practical Network Example, 171
5-39 Nonseries-Parallel Example, 175
5-40 Subgraphs of Example Network, 175
5-41 No. 5A Crossbar Network, 178
5-42 Lee Graph for Problem 9, 179
5-43 Network for Problem 12, 180

6-1 A Telephone Switching System Tree, 185
6-2 Telephonic vs. Commercial Processors, 186
6-3 Language Levels, 187
6-4 Assembly Language vs. POL, 188
6-5 ESS Processor Characteristics, 190
6-6 Microprogrammed vs Conventional Processors, 192
6-7 Reverse Polish Notation and Stacks, 194

8-1 Demand Curve, 216
8-2 Product Life Cycle, 217
8-3 Demand Curve Construction, 219
8-4 Value Curve Construction, 220

8-5 Bell System Cash Flow: 1981, 221
8-6 Project Cash Flow, 224
8-7 Accelerated Depreciation, 226

PREFACE

It has been the writer's observation for some time that the field of Telephone Switching has been strangely underrepresented in academe. Other aspects of Telephony such as Transmission are thoroughly covered in Engineering curricula, but a student fresh from school finds himself in a new world when he chooses an employer involved in Switching. Peculiar jargon, bizarre acronyms, new concepts beset the newcomer who had been confident of his grounding.

Telecommunications in the United States is on the brink of a major upheaval that will dismantle the most extraordinary company that has ever existed. Telephone service will become dependent upon the wares of a multiplicity of manufacturers, many of which will be new to the field. An understanding of Telephony in all its aspects is a goal devoutly to be wished of the future practitioners of the art. This book is an attempt to fill a portion of the need for such understanding at an introductory level.

The material for this book first began to be drawn together as a set of notes for a course that was taught for a group of some 250 Bell Laboratories Department Heads and Directors. It was then modified and taught at the advanced Undergrauate and Graduate level at the Illinois Institute of Technology, where the tempering effect of such exposure smoothed some of the rough edges.

The approach taken deserves some explanation. After a brief overview of Telephony, and a discusssion of basic principles, an historical progression of Bell System switching machines is described. The Bell machines are chosen because they predominate in the United States and have been extremely influencial in world machine architectures. Following one manufacturer's line of thinking usefully traces a continuity of philosophies, and the choice of the Bell machines provides an acquaintance with a large percentage of the world's machines. Then, as can be useful in other endeavors, comparative anatomy exercises become possible in the sense that significantly less detail on other manufacturers' products conveys their essence.

The important and peculiar topic of switches is covered, striving to convey to the student a degree of insight into their makeup and analysis, and traffic theory is introduced.

Stored program control, the soul of all modern telephone switching machines, is introduced as a distinct topic deserving special consideration.

The last topic, economics, was included because of the importance of monetary considerations in the field of Switching. A switching machine of moderate size represents millions of dollars of investment, and the tools of economic analysis are vital to the engineering choices that must be made.

The book is intended to be useful as a text for an introductory course in Telephone Switching at the advanced undergraduate and graduate level, and may prove useful as a self-study means for those for whom a college course is not available.

Much of the material in these pages represents the intellectual output of countless individual workers in the field of Telephony. I wish to express my appreciation for their contributions to the field. Unfortunately, they are so numerous as to be virtually unknowable. A few, however, have had direct impact upon this book, and I would like to acknowledge them specifically.

Amos Joel was the writer's mentor when this material was first presented within Bell Laboratories. His breadth and depth of knowledge in the field is virtually without peer, and his counsel was invaluable.

The writer is indebted to his supervision at Bell Laboratories, particularly Harvey Lehman, David Vlack, Eric Nussbaum, Warren Danielson and John Degan for their unflagging support in this endeavor.

The manuscript was prepared using the remarkably versatile phototypesetting system developed at Bell Laboratories. Judy Leth did an excellent job of keying-in the first draft, and Christopher Scussel was of invaluable assistance in the task of generating the phototypeset-ready output. Daniel Fyock's support in this activity is much appreciated.

The comments of students taking the course upon which this material is based were of remarkable value.

Frank Taylor and Walter Hayward made many useful suggestions, as did other reviewers within the Corporation.

A number of my associates were helpful in various ways, notably John Curtis, Larry Henderson, Bill Hoberecht, Kristin Kocan, Jim Leth, Lew Oberlander, David Spicer, and Jon Turner.

Finally, I wish to express my appreciation to my other colleagues who good-naturedly endured the preoccupation of their coworker while this text was in preparation.

B. E. B.

INTRODUCTION TO
TELEPHONE SWITCHING

CHAPTER 1

BASIC TELEPHONY IN BRIEF

1.1 INTRODUCTION

The function of establishing voice communication over a limited distance between a limited number of people via wire is simplicity itself. Only when the term "limited" is replaced by "virtually unlimited" do problems demanding high technology and involved theoretical considerations come into play.

The earliest practical telephone set transducer used was essentially a dynamic microphone that also served as the receiver; the user time-shared the device between his ear and mouth. Such a transducer severely limited the range of applicability of a system because of the feebleness of the induced signal. A dynamic microphone is basically a lossy transducer which converts incident acoustic power into an attenuated electrical signal. Even when maximum voice power is used, line losses rapidly take their toll upon the transmitted signal. Though such a transducer can be quite linear (an obvious advantage), linearity is a disadvantage with respect to low-level background noise, which it will faithfully reproduce.

The invention by Edison (and subsequently Blake) of the carbon microphone catapulted telephony into the realm of eminent practicability. It provided in the early years of the last quarter of the nineteenth century, the first practical, active circuit element. Not only was (and is) the carbon microphone capable of providing substantial power gain (of the order of 40 dB), but it is also a transducer of substantial dynamic range. It is not a perfectly linear transducer, but it is good enough! Indeed, its nonlinearity for low-level inputs tends to screen out background acoustic noise.

Even experienced electrical engineers are sometimes taken aback by the claim that the carbon microphone is an *active* device. Transistors, vacuum tubes, tunnel diodes, and the like are thought of as active beyond doubt, but attaching this adjective to a few grains of carbon apparently stretches credibility. The sticking point appears to be the dual role played by the carbon

1

microphone: transducer and gain producer. Unlike the transistor, which can amplify an incoming electrical signal into a larger output electrical signal, the carbon mike amplifies an incoming *acoustic* signal into a larger electrical signal.

A telling argument in this regard asks the question, "If a feedback circuit employing only a carbon microphone and passive elements (and, of course, a power source) can be demonstrated to sustain oscillation, must not the microphone be regarded as an active device?" Every engineer will immediately agree, but a remarkable proportion will insist on witnessing such a demonstration. (The design of such a demonstration is addressed in one of the exercises at the end of this chapter.)

Of importance almost equal to its performance is the inexpensive nature of the carbon microphone (cheap as dirt [or more properly, coal]). The active component is literally merely ground-up anthracite coal (chemically treated in modern times). The simplicity and low price of this transducer are the major reasons why it is still almost exclusively[1] employed for telephone microphones in spite of its disadvantages (which include a tendency for the carbon granules to *pack* and the variation of gain with bias current).

The receiver has remained basically unchanged except for the improved spatial efficiency due principally to improved magnetic materials. The *butter-stamp* receiver (a receiver shaped like the stamp used to impress the maker's symbol upon butter cakes) of yore was not designed for human factors considerations, but rather around the long magnet providing the biasing magnetic field.

Early transmission lines (emulating telegraphy practice) were uninsulated, open, single iron or steel wire with ground return (occasionally, a length of barbed wire fence was employed for a portion of the path). It was soon found that copper wire allowed longer transmission, and that the common ground return was a source of excessive noise. Copper, open wire paired line thus early became a standard that is still much in evidence (though copper-clad steel and steel are used in very long (e.g., 20 mile), sparse routes.

Powering of the telephone station was via per-customer battery in the earliest systems, and this presented a constant reliability problem. Eventually,

---

1. The Bell System C-36 handset used for long loops employs a receiver as the microphone abetted by a transistor amplifier, and electret microphones are coming into use.

a means[2] for powering from a common centralized battery without interference among lines was devised, and the reliability of telephone systems was substantially boosted. It was also recognized relatively early that the polarity of the dc potential applied to telephone lines had a dramatic effect upon their life (positive potentials promote corrosion in the presence of moisture), and negative potential became standard.

The alerting means rapidly advanced from a loud shout, to a *thumper* mechanism, to a two-gong ringer that has changed little in philosophy and form except for miniaturization.[3] Powering for the ringer was provided by a hand- or foot-operated magneto. Centralized ringing generators eventually became the standard.

Though small systems existed with no centralized switching means, the commercial offerings were almost immediately implemented with (manual) central switching facilities. This may well have been because the first such offering firm was the Holmes Burglar Alarm Company, which had experience in providing a central site, gathering burglar alarm lines from businesses and private homes. (Indeed, telephone service seemed a natural extension of the burglar alarm business, using a line for telephone purposes during the day and for burglar detection at night.)

Centralized switching by an operator relieved the user of the need to switch and performed the required functions well. To this day (and in the foreseeable future), there remains a residue of manual operator services that appear to be uneliminable via automatic equipment.

Centralization of switching also generated a need to inform the central office of the need for service (i.e., of a call *origination*). Early stations utilized the hand-cranked magneto,[4] which was vigorously spun to notify the operator of a desire for service. Later, dc current drawn by an off-hook phone would activate a *drop* at the operator position; still later, a board lamp; and yet later, one of several types of line *attending* means associated with automatic equipment.

---

2. The line-feed inductor.

3. Single gong ringers have been used for *loud* ringers and, in the Bell System, in the smaller station sets (P-ringers).

4. Still employed in simple military field telephones.

As lines grew longer and more numerous, and began to share poles with power lines, interference and attenuation began to become severe problems, and the probability of accidental *foreign* potentials and lightning strokes increased. The virtues of twisting wires and *frogging* were recognized, as was the need for balanced transmission lines to eliminate the effects of longitudinal noise. The use of heavier gauge wire for improved range was supplemented by the discovery of the merits of discrete *loading*, and carbon-block protectors against high potentials[5] began to be employed. Loading coils and protectors (carbon block or gas tube) are still employed by the millions, the former where needed, the latter virtually universally. *Heat coils*, which break the circuit if excessive current is drawn, are also often used.

Cables (groups of insulated conductors sharing a common sheath), used sparingly in early days (e.g., for river crossings) because of severe attenuation and crosstalk, have become common in above ground (aerial) as well as below ground (buried or in conduit) applications. Loading and careful manufacturing attention to twisting and pair placement have made the use of cable feasible.

Central offices grew in number, and means for communicating between them had to be provided. Shared wire pairs called *trunks* were employed, and operators at communicating offices agreed verbally over a voice path (*order wire*) upon which trunk to employ for completing a given customer connection. As telephone service spread geographically, a multiplicity of trunks served as segments of long-distance connections, and (eventually) the concept of a hierarchy of switching centers with distinct functions came into being.

Interest grew in coast-to-coast telephone communication, but the country's size proved a limitation. The largest practical wire gauge coupled with loading coils could not convey an electrical replica of the heartiest spoken word beyond about two-thirds the nation's breadth. Various schemes were tried, among the more interesting, using receiver-carbon microphone pairs as gain stages.[6]

Fortunately, the audion (the first vacuum tube) was invented at about this time, and amplifiers were designed that permitted coast-to-coast transmission. While solving many problems, amplification fostered others, e.g., singing, echo,

---

5. Protection means are usually provided at both the customer's premises and the central office.

6. This was a very reasonable attempted solution that failed because of mechanical resonances which, in multiple-stages, swamped out the intended signal. It is interesting to speculate how far this technique might have progressed if mechanical stagger-tuning had been employed.

and enhanced crosstalk, which are intrinsically not really soluble, but are thwartable. Such problems are compounded by the variety of possible connections, ranging from those between neighbors a few hundred feet from the switching office, involving no trunks, to those between customers on opposite coasts of the United States, involving up to nine trunks (two toll-connecting and seven intertoll).

As automatic equipment began to be used to perform switching functions, the signaling protocols between switching offices evolved, at first in accordance with the nature of the machines, later in accordance with the capabilities of the transmission medium and its environment. The functions of switching offices began to become somewhat specialized as they proliferated in a hierarchical and semihierarchical form.

The nature of switching equipment evolved as well, from distributed control to common control,[7] from gross motion to fine motion to no motion (electronic), from manual to electromechanical to electronic, from wired logic to stored program, etc.

Transmission techniques progressed from open-wire line to cable and radio carrier systems, including satellite links,[8] to digital (wire) carrier, and toward fiber-optic techniques.

1.2 THE TELEPHONE STATION

The telephone station has evolved considerably over the years, both in appearance and internal functional makeup. The pathologically simple combined microphone-receiver lasted only a very short time, supplanted by wall-mounted instruments, and, in turn, *candle-stick,* 200-, 300-, and 500-type sets in the United States. (Of course, the variety of telephone instruments in the world is wider, e.g., the *French-type* telephone, and many new housing types are being marketed.)

Besides physical design, improvements have included the hybrid (see Figure 1-1), the click-suppressor, automatic microphone gain control (as a function of

---

7. Distributed control is appearing again in the most modern switching equipment, which employs distributed microprocessors.

8. Which present special problems because of the transmission delay to and from the satellite. Such links are normally limited to one leg of the path.

loop length), and customer-adjusted bell control.

The hybrid transformer, which functions as a balanced bridge, theoretically would permit essentially complete separation of incoming and outgoing voice signals. In practice, the transformer is unbalanced to produce a controlled measure of *sidetone* (feedback into the earpiece of the customer's own articulations into the mouthpiece). This feedback is useful in regulating the customer's input volume, which he or she tends to adjust in accordance with individual perception of loudness as presented by the earpiece. The matching network uses lumped constant elements to approximate the impedance of the line. Half of the incoming and outgoing signal power is dissipated in the matching network, but the loss is more than compensated for by the human feedback, and by the impedance matching of the transmitter microphone (low Z) and receiver (high Z) to the line.

The click-suppressor is an amplitude-limiting device placed across the receiver to prevent extraordinarily loud signals from (potentially) causing ear damage. These are typically diode pairs or true varistors.

Microphone gain control is accomplished via shunting a large fraction of the available line current away from the microphone on short loops (where the lack of signal attenuation makes high gain even less attractive) and shunting virtually no current away on long loops (where less current is available and high gain is needed).

The ringer is constantly presented to the line, though it is ac-coupled via a capacitor. Its impedance is so high at voice frequencies that it is unnecessary to remove it during conversation. The device is mechanically and electrically tuned to the driving frequency (20 Hz) to increase its efficiency (which is, however, only about 1%), and to prevent it from responding to the rotary dial pulses (such unwanted response is called *bell-tap*).

The dialing mechanism is somewhat apart from the other telephone station functions; it consists of an escapement-timed progression of loop circuit openings (at a rate of 10 per second) for each rotary-dialed digit, or the production and transmission of a unique dual-tone, multi-frequency signal (see Chapter 4) in the case of (e.g.) a TOUCH-TONE™-dialed digit.

Depending upon the system, *talking battery* may be applied to the line from a point within the switching network, having removed the line circuit battery connection.

Access for testing the customer loop must also be provided.

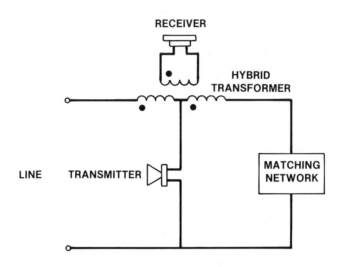

FIG. 1-1. The station set.

1.3 THE SWITCHING INTERFACE

Viewed toward the central switching machine serving a customer, the customer loop will normally join other loops in a fairly small exchange cable that will, in turn, join other such cables in a feeder cable, which enters the central office building in the *cable vault*. Cable is often pressurized with dry air at the central office to prevent intrusion of moisture through the cable sheath.

The customer pair proceeds up to the Main Distributing Frame (MDF), where protector blocks are installed for each line,[9] and where jumper wires are employed to associate (connect) the customer line with a network terminal pair. (In early systems, the network appearance of a customer line uniquely determined his directory number; in more modern systems, no such correspondence exists.)

---

9. In some systems a separate protector frame is employed.

Each customer loop is powered during talking by a (negative) 48-volt supply (nominal) which is ac decoupled (see Figure 1-2) via a line feed inductor that also serves as a split winding of a transformer which both dc isolates the network from the line and cancels longitudinal signals. The combination of twisted-pair lines and ac-balance armor the telephone circuit against much of the electromagnetic interference that would otherwise corrupt transmission. Successive twists in the wire-pair present oppositely poled areas to changing magnetic fluxes, reducing differential mode noise; and balancing permits virtually complete cancellation of common-mode noise. Surge-protection coupled inductors may also be employed to prevent damage due to surges not totally nullified by the protector block.

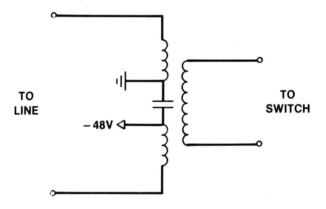

FIG. 1-2. The line interface.

The attending function was performed in early automatic systems by a sensitive relay, in later systems by saturable transformers, and still later by various electronic means.

1.4 TRANSMISSION

Telephonic transmission is a broad topic in its own right, and will be only touched upon here.

Subscriber Loop

The exchange cable customer loop facilities are chosen presently principally in accordance with *resistance* design:

> For most modern telephone sets alerting, supervision, and transmission all approach compromise when the dc resistance of the loop reaches about 1,300 ohms (assuming appropriate inductive loading for long loops). Clearly, a variety of combinations of wire gauges could meet such a criterion for a given loop, but for a limited number of available manufactured gauges, the most economical choice is usually unique.

For 26-gauge[10] cable, whose resistance per 1,000 feet at 20° C is 40.8 ohms, a 1,300-ohm loop is 15.9 kilofeet (just over 3 miles) long. Since the average Bell System customer loop is about 2 miles long, much of the loop needs can be satisfied with nonloaded 26-gauge cable (the smallest gauge manufactured in the Bell System for exchange cable application). When loops beyond distances reachable via resistance design must be served, *range extenders* are employed, which are capable of overcoming the long-loop limitations.

Interoffice Transmission

The nature of interoffice transmission facilities provided is a strong function of interoffice distances and of the position of the offices in the toll hierarchy. Transmission objectives advance in stringency from end-office communication through toll connecting trunks to intertoll facilities.

Trunks between end offices tend to be metallic[11] for short distances and digital carrier for longer distances. Toll connecting trunks are metallic, digital

---

10. American wire gauge.

11. Optical fiber systems are beginning to be used.

carrier (e.g., T1) or analog carrier (e.g., N), depending upon the distance. Intertoll trunks between distant offices use analog carrier (e.g., L) over coaxial cable, microwave radio, and more recently, high-rate digital over optical fiber.

1.5 THE SWITCHING MACHINE

The first switching entity was, of course, the operator, who reigned supreme until some 16 years after the telephone's invention.

The first automatic switching machines had to emulate the operator's task in spirit if not in detail. They had to encompass an addition to the telephone instrument (the dial) to permit communication in a form palatable to a machine. They had to provide means for communicating information to the customer without voice (e.g., *dial-tone* versus "Number please," *audible* versus "I'm ringing that number," *busy signal* versus "That number is busy"). They had to provide means for selecting a path and for setting up the path via closing switches as opposed to inserting a plug into a jack.[12]

As switching machines evolved, their capabilities expanded, allowing more economical use of equipment and offering more services to the customer. What began as the art of traffic engineering was eventually reduced to a science.

The proliferation of signaling languages, the need for charging capability, and the need to reduce trunking costs fostered tandem machines, and these needs on a larger scale bred the toll hierarchy.

Common control made it possible to perform "Gedanken"[13] path hunts through the network, increasing the probability of success or allowing less equipment to be used.

Electronics are beginning to reduce switching crosspoint size and price and permit an exchange of time division for space division.

EXERCISES

1. Design an experimental setup to demonstrate the active character of the carbon microphone. Consider either the use of a transformer or of a

---

12. Interestingly, at least one early invention employed cords and plugs that were automatically plugged into a jack field.

13. In thought rather than deed.

resistor and a capacitor (in addition to the transmitter, receiver and voltage source).

2. Assuming the telephone set resistance is 200 ohms and the line circuit resistance is 400 ohms, what emf is available at the telephone to operate the carbon microphone at the extreme of a resistance-designed loop? What current?

3. Consider two customers served by the same switching machine, both at the end of 1300-ohm loops. When they converse with each other, what gain must be provided by the carbon microphone to allow the sound power from the receiver to be equal to that incident upon the distant microphone, assuming losses through the switch are negligible? (The hybrid transformer in the station set loses about half the power on both transmit and receive.)

4. Determine which element in the station set dissipates the half of the incoming/outgoing power that is sacrificed by the use of the hybrid transformer.

5. Suppose that fiber guide were being considered for use in the customer loop plant. What would be its impact upon considerations such as powering of the station set, need for protectors, bandwidth available for other services, possibility of digital transmission, crosstalk, reliability, ease of maintenance, etc.?

READING LIST

There are many books on telephony of varying degrees of antiquity and quality; only a few will be listed here. More specific references are appended to appropriate chapters.

Early History

The Birth and Early Years of the Bell Telephone System, 1876-1880, R. Tosiello, Boston University Dissertation, 1971; available from University Microfilms, Ann Arbor, Michigan. (An incredibly detailed account of the problems and progress of telephony in its infancy.)

A History of Engineering and Science in the Bell System - The Early Years (1875-1925), M. Fagan, Editor, Bell Laboratories, Incorporated, 1975.

Transmission

Transmission Systems for Communications, Bell Telephone Laboratories, Incorporated, 1971.

CHAPTER 2

WHY SWITCH?

2.1 PRELIMINARIES

Telephony, in general, is a broader and deeper topic than it appears upon casual examination. Switching shares this property and, in addition, is obscured by an unintentional "smoke screen" of jargon that has grown up about the subject. Terms commonly used in other fields may have a different meaning in switching. For example, the term *network,* understood in electrical engineering to refer to a collection of resistors, capacitors, and inductors, refers instead in telephone switching parlance to a collection of switching elements arranged and controlled in such a way as to permit transmission between any of one set of terminals and any of another.[1] The term as used herein should also not be confused with a trunk network which interconnects a multiplicity of telephone offices (e.g., the Bell System network).

A number of unique terms have also been coined in switching. An example of this is the term *junctor* (which will be found in the glossary). Further, acronyms abound, sometimes pronounced phonetically (e.g., CAMA), sometimes spelled out (e.g., PABX). Still further, there are terms, especially associated with networks, that are named after early contributors to the field. As luck would have it, the names of several of them could be mistaken for acronyms. Examples of these are Erlang, Engset, Dimond, and Clos.

As is true of any nontrivial field, it is necessary to learn to speak the language, and a glossary has been appended to this text as an aid to that end. Most of the terms unique to telephony and used in this book are defined therein.

---

1. Strictly speaking, this definition is too broad, implying as it does the proper subset of all networks characterized as *full access.*

2.2 WHY SWITCH CENTRALLY?

Figure 2-1 illustrates the problem of attempting to provide telephone service to a multiplicity (n) of customers making use of an exhaustive network of connections. The number of 2-way lines required would be

$$\frac{n(n-1)}{2},$$

so that interconnection of, say 10,000 customers, would be somewhat impractical. Figure 2-2 illustrates that, in addition to many lines, a number $(n-1)$ of switching elements (crosspoints) would be required at each station[2] to permit selection of a line. (It thus becomes clear that switching must take place when more than two customers are to be served.)

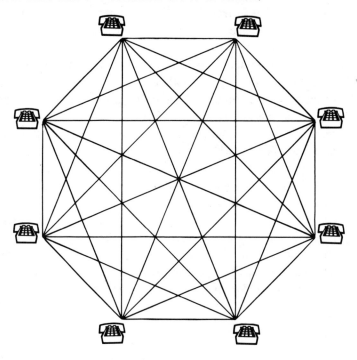

FIG. 2-1. The need for centralized switching.

---

2. Alias telephone.

It should be pointed out that the very early telephone switching systems used something rather like an exhaustive interconnection scheme. As the telephone became popular, however, the approach was soon recognized to be untenable; the beauties of centralized switching were thus recognized quite early. The advantages of centralized switching are illustrated in Figure 2-3, where only n lines are required to serve n stations. The interconnection pattern has merely shrunk in size; the number of interconnecting wires is the same (though they are much shorter), and the number of crosspoints is unchanged. However, a little thought reveals that the worst possible case occurs when one-half of the subscribers are talking with the other half. This fact is taken into account in Figure 2-4, where we see the number of switching crosspoints reduced by a factor of two, though any customer can still be connected to any other at any time (i.e., the network is *nonblocking*).

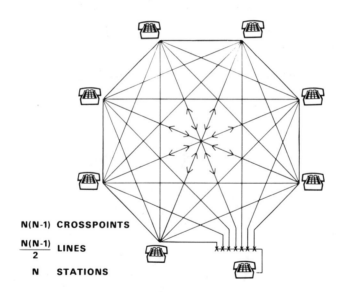

$N(N-1)$ CROSSPOINTS

$\dfrac{N(N-1)}{2}$ LINES

N STATIONS

FIG. 2-2. Non-centralized switching.

This number of crosspoints would, however, still remain unsatisfactory. It turns out in practice, as will be seen, that it is possible to design networks which use substantially fewer crosspoints than those indicated, when the statistics of network usage are taken into account and a small amount of blocking is permitted. Since only a fraction of the customers demand

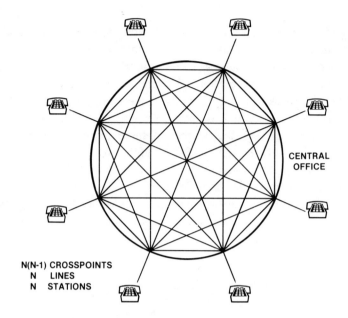

FIG. 2-3. Centralized switching: non-blocking, redundant.

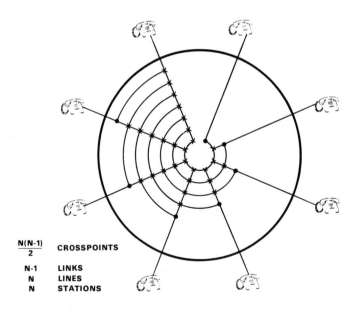

FIG. 2-4. Centralized switching: non-blocking.

interconnection at any one time, the network need be capable of simultaneously connecting only (about) that fraction[3] of the lines. Figure 2-5 illustrates a network that permits only one-half of the subscribers to simultaneously converse. The probability of blocking in such a network is calculable given the intensity and distribution of the traffic offered by the subscribers (such considerations are the subject of Chapter 5).

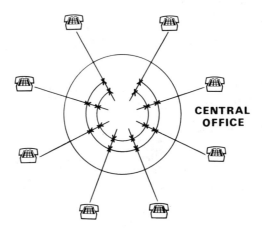

NL   CROSSPOINTS
 L   LINKS
 N   LINES
 N   STATIONS

FIG. 2-5. Centralized switching: blocking.

Now that the need to switch and to centralize has been established, the means whereby effective centralized switching can be accomplished may be addressed. The implications of centralized switching are:

1. Remote control of the switching operation is required because the switch is no longer in the hands of the customer,

---

3. Of the order of 10 percent.

2. The average length of the transmission paths increases (though they are fewer in number) because the office[4] is centrally located, and must be traversed by every call.

A simplified diagram of an automatic switching machine is shown in Figure 2-6, where the disposition of the lines served by the machine and the trunks to and from other offices is depicted.

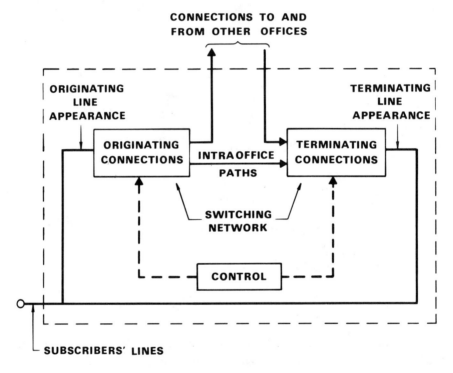

FIG. 2-6. Simplified automatic switching machine.

The diagram is intended to characterize the activities of a switching machine but is truly descriptive of few actual machine types.

The customer lines are shown to have two appearances on the network, one an *originating* appearance, the other a *terminating* appearance. In many

---

4. The term *office* or *central office* is used somewhat loosely in telephony. It is correctly synonymous with *entity*: a switching network and its control. It is, however, frequently used to mean *office code*; several offices codes (also often called *NXXs*) are typically served by a single entity. A *theoretical office* is a number group given separate rate treatment.

switching machines, these appearances are coalesced into a single appearance by, in effect, *folding* the network about its centerpoint. Some networks are folded with respect to one component of traffic (e.g., intraoffice traffic) but not with respect to another component (e.g., interoffice traffic).

Connections to and from other offices are via *trunks,* which may be metallic (a wire-pair per trunk) or some form of multiplexed carrier system (e.g., T1, a PCM system). In general, trunks are thought of as expensive facilities compared to lines. They may be quite long, and their terminals may be relatively complex. Further, because loss of a trunk may seriously cripple the communication between offices, testing is frequent and rigorous.

The intraoffice paths indicated may be called network junctors or intraoffice trunks, depending upon the machine, the points of connection upon the network, and the manner of their handling.

The control as depicted implies a centralized intelligence (which, indeed, characterizes many modern systems), but such implication is not intended; the earliest and still most pervasive system employs a finely distributed control. Other systems, while more highly centralizing control, split it between an originating and a terminating portion of the machine, which are treated as virtually independent. A multiplicity of pooled controllers is also employed in some systems.

In general, when a subscriber originates a call, receives dial tone (indicating connection at the central office of a dial-pulse or TOUCH-TONE reception means), and dials a complete number corresponding to another line served by that office, a path is set up through the originating network, an intraoffice path, and the terminating network to the called line. If the call were to a line served by another office, only the originating network would be traversed, followed by an interoffice trunk (possibly several in tandem, as will be described later), then the terminating network in the destination office. Similarly, an incoming call from another office traverses only the terminating portion.

2.3 THE ANATOMY OF A SIMPLE CALL

Consider the functions (see Figure 2-7) that must be performed by this now centralized switching machine or *central office* in handling a simple call. When a customer's telephone goes off-hook (originates), he in effect signals the office that he desires service (signaling will be considered later as a separate category). A centralized switching machine must be capable of sensing that the customer requires service; monitoring for such an event is referred to as *attending.* Attending is particularly important because it is implemented, in general, on a per-line basis, so that the amount of equipment involved is in one-to-one correspondence with the number of lines and, therefore, can potentially represent a substantial cost.

CALLER	CALLED	CENTRAL OFFICE
		ATTEND
OFF-HOOK (ORIGINATION)		
		CONNECT SIGNAL RECEPTION MEANS AND RETURN DIAL TONE
DIAL		
		STORE INFORMATION, TRANSLATE DIALED NUMBER TO EQUIPMENT LOCATION AND FIND PATH; TEST FOR BUSY
	BUSY	
		RETURN BUSY SIGNAL AND SUPERVISE UNTIL ONHOOK
	IDLE	
		ALERT (RING), SUPERVISE RETURN AUDIBLE RINGING
	OFF-HOOK	
		CUT THROUGH TALKING PATH AND SUPERVISE
ON-HOOK		
		TEAR DOWN PATH AND CHARGE

FIG. 2-7. Anatomy of a simple intraoffice call.

Once the central office has recognized that an origination has taken place (by noting the off-hook status of a given line), it is necessary that it connect to the line some means for receiving information and notify the customer (via dial tone) that the office is ready to receive the information. Assuming that the off-hook station is not a manual line, information is entered by the customer, using the dial, and is received and recorded (in some sense) at the central office. This information is then interpreted by the central office equipment to identify the locale of the called line (within the same office or elsewhere). Assuming that the called line is within the same office (i.e., the call is intraoffice), a busy test is made of the called line and, if idle, it is rung (*alerted*). The called line is *supervised,* awaiting answer, and when (and if) answered, a talking path is established. The talking path is then supervised until one or the other party goes on-hook, whereupon that path is taken down.

If the called party were not served by the same office (an interoffice call), a search would be made for an idle direct trunk to the office serving the called line or to an office higher in the hierarchy that would be able to further the progress of the call toward its destination. When the distant office is ready (in some cases,[5] whether or not it is ready) to receive information about the called number (only the last four *directory number* digits if the trunk is direct; more or all if not), it is transmitted (*signaled*).

When ringing takes place, a muted version of the signal (*audible*) is supplied to the caller until the call is answered or abandoned. If the called party's line is busy, or the call is blocked in the network, or the necessary trunks (for an interoffice call) are all busy, the customer is informed via a busy signal in the first case and *fast busy*[6] or *reorder* in the latter two cases.

Depending upon the switching machine type handling the call, the above functions are performed in remarkably different ways. The next chapter describes various means devised to perform the functions of a switching office. However, before learning the details of existing and proposed switching machines, which may obscure the underlying principles, it is helpful to figuratively pour alcohol over the whole of switching technology to obtain a "tincture of switching," and the next section (2-4) will so endeavor.

The basic task of providing the kind of telephone service commonly available 30 or 40 years ago is fairly simple. This class of service, commonly known as POTS (Plain Old Telephone Service), has been augmented in modern times by many special services, some of which are very complex and demanding in their requirements upon the capabilities of the switching machine which provides them. Though a few of these will be mentioned specifically in the following pages, they will not be considered in great depth. The flavor and spirit of switching are reasonably well conveyed by an understanding of POTS, but it should be kept in mind that the system architecture of modern switching machines is heavily influenced by the need to provide the new services.

---

5. Specifically, when the originating entity is a step-by-step machine.
6. Fast busy or reorder is similar to simple busy, but the tone is interrupted at twice the rate.

2.4 SWITCHING PRINCIPLES

From the foregoing, it is clear that the telephone switching machine must perform:

1. Signal reception.

2. Signal interpretation.

3. Storage (buffering and translation).

4. Path selection.

5. Network path provision and control.

6. Signal transmission.

The principles of telephone switching are essentially spelled out in the definitions of these vital functions.

Signal reception means must be provided for signals which may be received from customer lines or from trunks (and possibly other "private" interoffice signaling paths, e.g., CCIS,[7] described in Chapter 4) and may be in the form of DC loop closures, TOUCH-TONE, Single Frequency (SF), Multi-Frequency (MF, 2-out-of-5), and others. Usually, these means consist of a dedicated portion that detects a need for service, and a shared portion that is commonly only momentarily associated with a signaling entity. Often the dedicated portion must be switched off the path after its detecting function has been performed, because the loading effect of its presence is undesirable during the remainder of the signaling process or after a speech path is established.

Signal interpretation is the most complex of the switching machine functions (particularly in the more modern common-control machines). It is in the interpretation process that the idiosyncrasies of the human signaler must be dealt with, as well as the multitude of signaling languages, and the broad regime of the switching plan consulted and its discipline invoked. Implemented in a distributed form in simple, early systems, the interpretation means is more concentrated in electromechanical common-control systems and even more concentrated in ESS systems. The equipment is time-shared in all three cases:

---

7. Common Channel Interoffice Signaling.

the first (early systems), for the duration of the call; the second (electromechanical common control), for a few 100-millisecond intervals during call setup; and the third (ESS), for a few 100-microsecond intervals during call setup.

Storage is necessary primarily for buffering of signaled information, retention of call status, and as a repository of translation information (the additional uses for memory in ESS machines [e.g., program storage] are not primary requisites for telephone switching). Signals are frequently transmitted in serial form; and though in the simpler systems they are digested or relayed serially, more typically they are stored temporarily in parallel form for examination and digestion or retransmission. Call status information must be retained in some form (call progress activity may be viewed as the functioning of a classical finite state or sequential machine, requiring memory by definition). Translation information, providing correspondences between directory numbers and equipment addresses and vice-versa, area and office codes and appropriate trunk group addresses, etc., though relatively static, must still be stored in some form.

Path selection through the network consists of testing appropriate portions of the network until an idle path is found between the terminals of interest. The "appropriate" portions are virtually autonomously identified in the early systems but must be "computed" in common-control machines via wired logic (in the electromechanical systems) or stored program (in ESS). The busy/idle status of a network element is testable at the network in electromechanical systems but is retained in a *network map* in memory in, e.g., Bell System ESSs.

Network paths and their control must be provided in some sense. In the early systems the path selection was[8] autonomous, and control direct via the customer (even from a distant office). In later systems, these functions are performed by control common to many paths. There exists an approach (implemented, for example, in the Morris, Illinois field trial with gas tubes) known as *end marking,* whereby a network is provided of such nature that applying a potential to the terminals between which such a path is desired, autonomously brings such a path into being. No sufficiently robust and

---

8. And is: the "early" systems still exist in large numbers.

economical network with this property has yet been devised for central office use.[9]

Signal transmission is necessary for communication to the customer and to distant offices. In the early systems, the former function (signaling the customer) was performed at the appropriate level in the switch train, the latter (signaling a distant office) via routing the customer dial pulses out onto a trunk. In later systems, the signal-generating means are time-shared and switched-in when needed (these means are multiple and must be searched amongst for an idle member).

An additional important function that must be performed by a switching machine is that of interfacing electrically with the hostile and demanding outside world. This function may be characterized by the tasks required of the line interface circuit, whose role has been dubbed that of the *BORSHT Circuit*,[10] which is an acronym for:

Battery:　　　　A source of EMF (usually -48 volts) must be provided to the line to allow detection of station status and dial pulses (or to power a tone signaling pad), and to power the carbon microphone.

Overvoltage:　　Extreme voltages impressed upon an incoming line by a lightning stroke or power line cross must be attenuated to safe levels without impairing transmission under normal conditions.

Ringing:　　　　The 20-Hz, 88-volt RMS ringing signal standard in the USA, or its counterpart elsewhere, must be applied when required.

Supervision:　　The status of a line (on-hook or off-hook) must be monitored by sensing the current drawn. This task becomes difficult when lines are very long (requiring high sensitivity) and when they are made leaky by wet weather (requiring low

---

9. With the exception of a system developed by ITT (the TCS5).

10. This mnemonic is traceable to one J. E. Iwersen of Bell Telephone Laboratories, and may be puzzling to some readers. The Borscht Circuit is the name given by entertainers to the theatrical circuit comprising resorts in the Catskill Mountains frequented by patrons presumably fond of borscht.

sensitivity).

Hybrid: The hybrid circuit separates the incoming and outgoing signals delivered to or issuing from, respectively, the receiver or transmitter.

Testing: Metallic access to a line is required for the testing that takes place periodically.

For digital switching systems, the acronym becomes BORSCHT, where the C stands for Coding (i.e., A/D and D/A conversion).

2.5 REAL-TIME INTERACTIVE OPERATION

The terms *real-time processing* (operation at a rate great enough to keep pace with fairly rapidly changing events), and *interactive* (having the capability of interacting with a human in accepting information, processing it, and responding in an intelligible manner on a multiple-transaction basis), are often presumed to apply only to very modern computer systems. A little reflection will reveal, however, that these attributes can properly be applied to telephone switching machines 75 years old or more.

Even very early electromechanical equipment was capable of speeds permitting a form of real-time operation: machine response to a customer's inputs appeared to be virtually instantaneous. Interactive capability was also provided very early: the customer's input repertoire included off-hook, on-hook, dialing, flashing, and permanent signal; machine responses included dial tone, dial tone removal, line busy, equipment busy, audible ring and its removal, and howler application. The processing performed is beyond the consciousness of the customer, who is made aware only of the net results of each transaction.

The peculiarities of dealing with humans, recognized now as an important consideration in real-time, interactive computer system software design, had to be wrestled with much earlier in telephony. The main stem of an algorithm for serving a perfect customer had to be encrusted with alternative procedures to accommodate arbitrarily irrational customer behavior; all conceivable possibilities had to be considered and dealt with in a consistent manner, yielding a result that would not confound or confuse a layman user. These requirements remain today, multiplied by the increased breadth of services provided and planned. Tandem and toll offices, since they deal with other offices (which are more civilized in their behavior than humans), have somewhat simpler call-handling algorithms.

2.6 SWITCHING PRACTICES

Attempting similarly to extract the essence of switching practices, the following general observations on the application of switching principles may be made.

Switching machines have historically been designed and constructed with the view that they may be expected to provide service for 40 years or more. In general, they are installed with sufficient capacity for the initial application plus the expected growth for several years. In most cases, augmentation of the initial capacity is expected to occur periodically, and the machines are designed with expansion capability.

Every reasonable effort is made to avoid failures and to correct them rapidly when they occur. Preventive maintenance is practiced where applicable, and many offices have one or more maintenance people (craftspeople) on duty at all times.[11]

Backup power in the form of storage batteries (in all central offices) and engine-driven generators (in the larger installations) relieve the offices of immediate dependence upon externally provided power.

Subscriber lines and trunks to other offices typically enter the central office building in the basement through the *cable vault,* where they leave the sheathing of the cables (many of which are pressurized to prevent moisture entry) and proceed up to the *Main Distributing Frame* (MDF), where they are separately connected (see Figure 2-8). The MDF provides the interface between the outside world and the switching machine in that it houses the *protector blocks*[12] and physically supports the jumper wires that electrically associate lines and trunks with terminals upon the switching network. In general, the protector blocks consist of shunt carbon blocks[13] and series heat coils, the former to limit the potentials allowed to reach the network to about 600 volts (lightning strokes and contact with high-tension power lines occur occasionally), the latter to limit (by opening the circuit) the current drawn

---

11. Although much progress has been made in centralizing monitoring facilities for several offices and dispatching maintenance personnel when required.

12. Modern machines often have a separate *protector frame* which precedes the MDF proper.

13. Gas tubes with lower breakdown voltages are also used.

when some misadventure befalls the loop. (It should be mentioned in passing that the electrical rigors experienced on occasion by lines and trunks are among the chief factors which historically militated against application of solid-state devices in the line circuits and networks.)

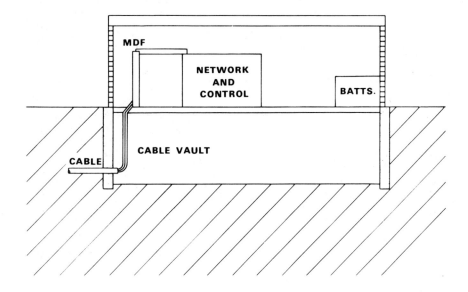

FIG. 2-8. Cross-section of a small office.

Connection thus made to the network terminals completes the capability for telephone switching in its most basic form. Line and trunk circuits contain the means for detecting originations or *seizures,* respectively, and for some switching of equipment onto the line. The line circuits, being so numerous, must perforce be simple, while the trunk circuits are often quite complex, with built-in timing means, etc.

By way of example, there are about 10,000 central offices (more than 15,000 central office codes) in the Bell System, though most of them are quite small (see Figure 2-9). The large ones, however, are very large indeed, so that more than 75 percent of the lines are served by less than 25 percent of the central offices. The pace of activity varies from the typically rural *Community Dial Office* (CDO) serving perhaps 200 lines, which is unmanned though visited once or twice a week by a "jack-of-all-trades" maintenance person, to the big city *metropolitan wire center,* which might contain several 20,000 line offices, each heavily and constantly manned, with personnel installing additional equipment on a nearly full-time basis and the maintenance personnel specialized in their duties. Figure 2-10 details the distribution of machine types and numbers of lines served.

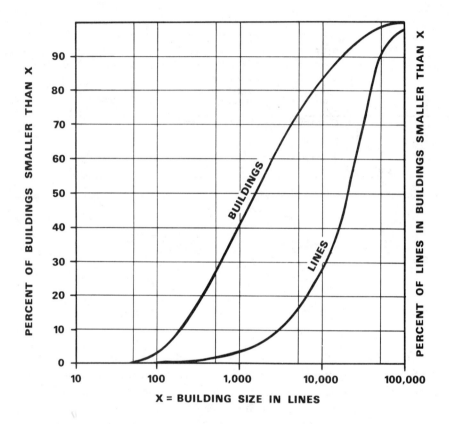

FIG. 2-9. Cumulative distribution of local exchange buildings and lines (est.).

Switching equipment has historically been designed to operate satisfactorily without forced-air cooling or air conditioning; cooling means were provided only for the comfort of personnel. The needs of modern equipment are beginning to change these philosophies.

MACHINES	TYPE	LINES
No.1 ESS No.2 ESS No.2B ESS No.3 ESS No.1A ESS No.5 ESS	Electronic	44M
SXS X-Bar	Electromechanical	41M

FIG. 2-10. Distribution of machine type usage.

EXERCISES

1. Alerting usually entails ringing a bell in the subset, which requires that the switching network be capable of transmitting substantial power. How else might the alerting (ringing) function be performed with equal effectiveness but with a less stringent requirement upon the power-handling capability of the network?

2. Would providing some of the power for subset operation from the customer's premises be a reasonable approach? What advantages or disadvantages would this have?

3. Devise a circuit to be used for the attending function which will load the line so little that it will not be necessary to remove it from the line after an origination is detected?

4. Power companies share their generating capacity to aid in providing power for peak load times which follow sequentially across the time zones. Would a similar approach be reasonable for telephone traffic-handling capacity?

5. What new services can you think of that could be offered by a telephone company to its subscribers and would be useful or popular enough to be economically viable?

READING LIST

Textbooks:
There are remarkably few textbooks that can be recommended without reservation to the newcomer to telephone switching. Many such texts are written by authors not associated with any telephonic administration or equipment manufacturer, and tend to generalize to include all systems, thereby sometimes distorting the treatment relative to any one system.

Basic Telephone Switching Systems, 2nd ed. D. Talley, Hayden Book Company, 1979. (A good introduction to the subject.)

Switching Systems, AT&T, 1961. (An authoritative but somewhat dated treatment.)

A Time For Innovation, A. A. Collins and R. D. Peterson, Merle Collins Foundation, 1973. (Quite up-to-date philosophically, but rather opinionated.)

Telecommunication Switching Principles, M. T. Hills, MIT Press, 1979. (A well-done British view.)

Communication Switching Systems, M. Rubin and C. E. Haller, Reinhold Publishing Corporation, 1966. (Much extraneous information, but good browsing.)

Automatic Telephone Practice, H. E. Hershey, Technical Publications, 1946. (This is a down-to-earth treatise on the switching systems of the independent telephone industry to the end of World War II, but much more, it is interspersed with the "swan song" of a gifted telephone engineer, frankly describing the events and personalities that shaped the "other" major telephone presence in the U.S.)

CHAPTER 3

CURRENT SYSTEMS

Current systems may be classified on the basis of their function or implementation as *local* and *tandem, toll,* or *operator.*

3.1 LOCAL AND TANDEM

In the Bell System, local and tandem[1] switching are handled by Step-by-Step (SXS), Panel, Crossbar (X-bar), and Electronic Switching Systems (ESS).

Tandem offices provide an intermediate switching point between local offices, reducing the number of direct trunks required (see Figure 3-1). The function of a tandem office is similar to that of a toll office, but is usually 2-wire rather than 4-wire because of the short haul nature of the traffic carried (i.e., transmission requirements are less stringent). Some tandem capability can be provided in otherwise *end offices* as well. A tandem office does not necessarily have trunks to every local office in its area, and local offices will often have direct trunks between them as well as trunks to a tandem office.

Step-by-Step and Panel systems came into use in the Bell System about the same time (c. 1919). SXS continued to prosper until quite recently, while manufacture of new Panel offices ceased in the 1940s as Crossbar systems became available. Electronic Switching Systems are the most recent addition. As will be seen, the systems are quite different in form, internal function, range of application, advantages, and disadvantages.

---

1. Pairing these topics, though customary, is somewhat artificial in the sense that local tandem machines are functionally linked to toll machines in that their switching features are virtually identical. However, the primary function of local tandem machines is the switching of interlocal traffic, and they have no standing in the toll hierarchy. The term "tandem" is also applied to certain toll activities; as used herein, *local* tandem is implied.

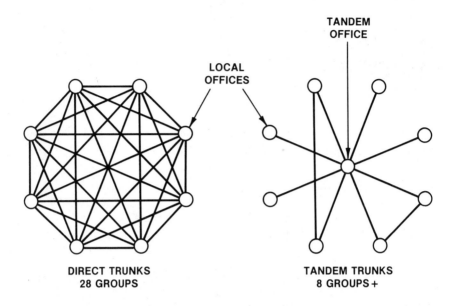

FIG. 3-1 The tandem office.

Step-by-Step (SXS)

Step-by-Step was invented in its original form by an undertaker and embalmer by the name of Almon B. Strowger in 1889, in Kansas City. As the story is told, there was only one competitor to Mr. Strowger in his part of town, and when that competitor's sister-in-law became the local exchange's switchboard operator, Strowger noticed his business falling off. Enraged, he set about inventing an automatic system which would replace telephone operators. The original switch differs relatively little in concept from those in use today.

The first commercial installation of SXS was in 1892; the first Bell System installation of a complete central office was in 1919 (though a number of these machines were already in the Bell System as the result of acquisitions of independent companies).

SXS has proven a remarkably versatile, rugged, and economical system. It is the only system, for example, that has been used in the full gamut of community dial, local, tandem, and toll applications, and is capable of use in office sizes ranging from below 100 lines up to 40,000 lines (and, theoretically, beyond). Further, a fortunate result of the distributed nature of the equipment is that an office rarely becomes totally inoperative, so that the level of reliability to which customers using this equipment became accustomed set a standard which the most modern equipment is hard-pressed to meet.

A SXS system makes use of various forms of a two-dimensional (cylindrical coordinate with constant *r*) stepping switch which is characterized as a *gross-motion* switch (see Figures 3-2 through 3-4). Briefly, such a switch can be used to respond to a (rotary) dialed digit by stepping vertically a number of levels equal to the number of dial pulses received, then (in the case of *selector* switches) searching horizontally for an idle link to the next switch, which will respond to the next digit dialed. Other types (*connector* switches) can respond to two successive dialed digits, responding to the second digit by stepping horizontally a number of positions equal to the number of dial pulses. Still others can be "turned around" and used as *line finders*, searching both vertically and horizontally for an originating line.

Figure 3-2 depicts the terminal bank. The central shaft (shown truncated) is capable of vertical and rotational motion. It will be noted that there are three contacts, two (tip and ring) upon the lower wiper arm (making up the voice path), and one (the sleeve) upon the upper wiper arm. The sleeve lead is used to provide the current to "hold up" the path through the network; it also serves the useful purpose of indicating busy/idle status to other switches searching for a path.

Figure 3-3 displays a switch with the dust-cover removed, showing the mechanism that propels the shaft upward and around, and the relays used for timing and control.

Figure 3-4 shows a typical office installation. The equipment bays are 11 feet high, necessitating the use of ladders for maintenance.

The talking path is held up by successive switches corresponding to the hierarchy of dialed digits, and control is passed successively from switch to switch. SXS systems are therefore classified as *direct progressive* because control of the network is direct (by the customer) and the call advances in a sequential, progressive manner through the network.

Figure 3-5 depicts a simplified case of a 10,000-line office. The calling line going off-hook causes an idle line finder to advance to the level and position of the line appearance and return dial tone. The caller dials a 3, which causes the first selector to advance up to the third level and search across for an idle link to the next selector. A dialed 5 causes the second selector to search on its fifth level. The dialed 7 causes the connector to advance to its seventh level, and the dialed 9 causes it to advance across to the ninth position, test for busy, then apply ringing.

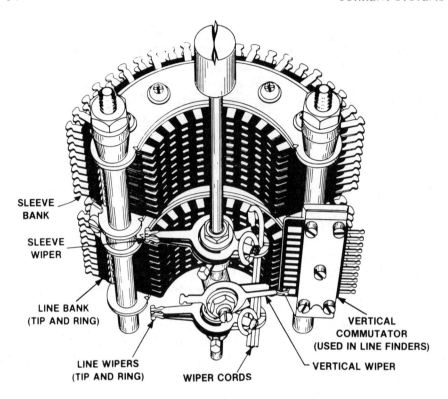

SLEEVE
BANK

SLEEVE
WIPER

LINE BANK
(TIP AND RING)

VERTICAL
COMMUTATOR
(USED IN LINE FINDERS)

VERTICAL WIPER

LINE WIPERS
(TIP AND RING)   WIPER CORDS

FIG. 3-2. SXS terminal bank.

FIG. 3-3. SXS switches.

FIG. 3-4. SXS switch bay.

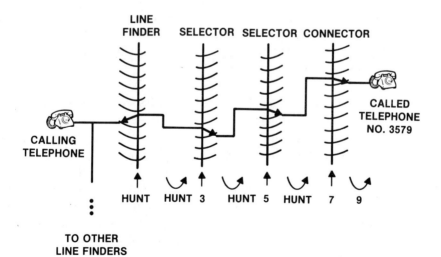

FIG. 3-5. A call in a 10k SXS office.

A variation upon the basic SXS system provides means for collecting dialed digits in a register, then spilling them forward into the switchtrain. This has the advantage of the possibility of retrial when no idle path can be found, and the philosophy can embrace TOUCH-TONE dialing. This control technique is classified as *register progressive*.

There has been interest for many years in phasing out SXS because of maintenance costs and inflexibility with respect to new features (features such as TOUCH-TONE are accommodated with difficulty, and offerings such as *abbreviated dialing* are virtually impossible). However, SXS continued to be manufactured in substantial quantities for some time, and only recently has begun a down-turn.

SXS has displayed such robustness chiefly because it is so inexpensive in small installations, and first cost has been an important consideration among capital-limited operating companies.

SXS must, of course, be effectively interfaced with by modern systems because of its present and long-term ubiquity. An example of the difficulties that an interface between a 1920s system and a 1980s system can produce is the manner in which an interoffice call is handled by SXS:

> When the first three digits (the office code) are recognized by a SXS end office with a direct trunk to the called office, a search will be made for an idle outgoing DC trunk (SXS dial-pulse trunk) to that office; when such an idle trunk is found, it is seized. Meanwhile, the dialing customer has begun dialing his fourth digit, which will be directed, along with the final three digits, over the trunk to the destination office.

> If the destination office is served by an ESS machine, the incoming trunks (and lines) are not monitored (attended) constantly, but are scanned periodically. If the chosen trunk were near the end of the trunks searched amongst by the originating (SXS) office, so that the search consumed considerable time, and if the originating customer were a speedy dialer, the first pulse of the fourth dialed digit would appear almost immediately after trunk seizure. If the scanning rate of the ESS office were not high enough, the pulse would be lost, the fourth digit would be interpreted as one less than its real value, and an incorrect connection would be made.

In fact, the scanning rate required for DC incoming trunks from SXS offices is among the highest required for any purpose in an ESS office. (The same problem exists in Crossbar offices which provide special *bylinks* to deal with incoming trunks from SXS offices; the term "bylink trunk" has thus crept into the vernacular.)

About one-third of the Bell System lines are served by some 6,000 SXS machines.

Panel

At about the same time that the Bell System first used SXS, it began installing its first Panel systems (first commercial installation in 1921). While archaic in appearance by today's standards, these systems performed yeoman's service in high-traffic metropolitan areas. They were actively being installed over a period of only about 20 years, but their retirement (until recently) was very slow. They were rapidly supplanted by ESS machines, and had disappeared by 1982.

Except for historical interest, the usefulness of studying such a system might be questioned. However, it provides another example of a system with an unusual signaling language with which ESS and other systems have had to interface, and its philosophies influenced later systems.

Briefly, a Panel switch is categorized as gross motion and progressive, with indirect control. The "grossness" of its motion surpasses that of all other telephone switches, accommodating one-dimensional movements measurable in feet (see Figure 3-6). The power for the switch rod (30 of which can be seen on the front of the frame, and 30 more of which are on the back) movement was supplied by electric motors at the bottom of the equipment bays, which were in constant motion, driving horizontal shafts that could, in turn, provide vertical motion to selected switch rods via cork clutches. The effect to the viewer was vaguely reminiscent of an old-time shoe repair shop with its multiplicity of belts deriving power from a single drive shaft.

The switch provided access to 500 terminals via selection of one of five sets of brushes on a switch rod (each with access to 100 terminals), thereby reducing (in comparison to SXS) the number of switching stages required.

The Panel machine was the first to employ common (centralized) control vis-a-vis direct control.

The Panel system structure is indicated in Figure 3-7. The system was organized into two virtually independent portions, originating and terminating. The bulk of the intelligence of the machine resided in the *senders*, which

FIG. 3-6. Panel bay.

remained associated with a call until a complete path was set up. They made up a pool any of whose idle members were candidates for handling the next call. They, in turn, were clients for a pool of *decoders,* which contained information relating the dialed office code to the proper corresponding trunk group (inter- or intraoffice), in terms of the necessary selector frames and levels capable of accessing it.

Figure 3-8 depicts in more detail the structure of a Panel machine. The line finder frame was organized into 10 banks of 40 lines each, and the final selector frame was organized into five banks of 100 lines each. These frames corresponded in function to the SXS line finder and connector, respectively.

The district, office, and incoming selector frames were essentially identical to the final frame (five banks of 100 terminal triplets each) and corresponded to SXS selector stages.

In the Panel machine, however, the selectors did not respond directly to the customer's dialed digits, but to signals from a sender. A sender remained associated with a call until the necessary path was set up from the calling to the called line. It controlled path set up not only in its own machine's originating and terminating networks, but in the terminating network of a

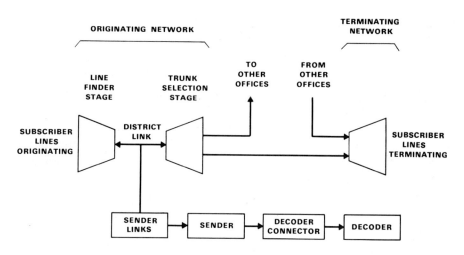

FIG. 3-7. Panel organization.

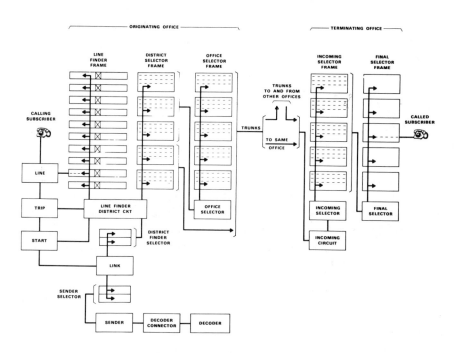

FIG. 3-8. Panel office detail.

distant Panel office to which a call was being placed over a trunk.

The controlling signals were in a form known as *revertive pulsing,* which is a feedback technique analogous to the situation where one person fills a glass until the other person says "when." A *start pulse* from the sender caused the recipient frame to send back a sequence of pulses that were compared by the sender to the number to be conveyed; when a matching number of pulses had been received by the sender, a *stop pulse* was sent. The pulses sent back by the recipient frame were produced by a commutator that was swept by a brush as the vertical rod was propelled upward; thus the position of the rod was fed back to the sender until it was satisfied.

When a customer's telephone went off-hook, the relay associated with his line activated a *trip* and *start circuit* that caused a line finder rod to rise and connect the line to an idle *district circuit.* At the same time, an idle sender was found, connected to the line, and dial tone was returned to the customer. After the office code was dialed, the sender requested connection to an idle decoder, which provided routing information and then disengaged. After the remaining digits had been dialed, the sender caused a rod on the district selector frame to advance to a level on the selected bank corresponding to the first of a group of trunks or of links to the office selector frame. The rod was advanced through the group until an idle terminal was found. Where necessary, a similar process was carried out in the office selector frame.

When a trunk was selected, the incoming circuit in the terminating portion of the office (or distant office) conditioned the network to respond to the (revertive) control pulses from the sender in the originating portion of the office. A rod on the incoming selector frame advanced up to the proper level, tripped the brush on the proper bank and searched for an idle link to the proper final selector frame. A final selector frame rod advanced to the level corresponding to the called customer's *tens group,* then advanced at a slower rate to a terminal corresponding to the customer's units position. Tripping the brush in the proper bank completed selection of the called customer's terminal, which was then supplied ringing, or, if busy was detected, busy tone was returned to the calling customer. The sender released after delivering information on the last dialed digit.

In a 10,000-line office, the incoming frame would have each bank divided into 4 link-terminal groups, so that choice of a bank (out of 5) selected to within 2,000 lines, and that of a level on the bank selected to within 500 lines. Similarly, for the final selector frame, choice of a bank selected to within 100 lines, and successive movements on the bank selected to within 10 lines, and finally the line itself.

> Note: As was true for SXS systems, a line's directory number uniquely determined the physical placement of its appearance on the network. Such a requirement was an encumbrance, and is eliminated in subsequent systems.

The sequence of operations performed by the Panel machine selector rods was controlled by rotary electrical cams known as sequence switches, one per rod, which were not unlike those used to control timing in a modern washing machine. Each switch comprised 26 disks with 18 discrete positions.[2]

The Panel system design was aimed at efficient handling of interoffice traffic (which predominates in the urban metropolitan areas where this machine was primarily employed) at the expense of intraoffice traffic.

Though withdrawal was completed by 1982, there were about 40 of these machines still in service as late as January, 1977, serving 0.5 million lines. Anyone interested in the heritage of telephony is encouraged to avail himself of the opportunity to view the Panel system now on display at the Smithsonian Institute in Washington, D. C.

Crossbar

The Crossbar network concept originated in the Bell System in 1913, but failed to be developed. The idea was pursued in Sweden, however, and was applied commercially. Eventually, the merits of the Crossbar structure were rerecognized in the United States, and a series of systems employing this switch were developed.

In order of appearance, these systems were:

---

2. The modern-day reader might best visualize the sequence switches as functional equivalents to a microprogram, with each disk corresponding to a microinstruction bit position, and the presence or absence of a conductor at a given peripheral position on a disk corresponding to the respective bit value, the net result being a series of control pulses applied in parallel to a number of control leads.

No. 1 XBAR   - 1938
XBAR TANDEM - 1941
No. 4 XBAR   - 1943
No. 5 XBAR   - 1948
No. 4A XBAR   - 1952
No. 5A XBAR   - 1972
No. 3 XBAR   - 1974.

It will be noted that one number is missing, and one is not in chronological order. Originally, two missing numbers were associated with versions that did not reach development, or were already preempted by other pieces of equipment; one of these numbers (No. 3) was finally employed some 30 years out of sequence because it had originally been associated with another system type.

The nature of the Crossbar switch and its control philosophy will be described in association with its first application in the United States: the Bell System's No. 1 Crossbar.

No. 1 Crossbar

The limited flexibility and high maintenance costs of the Panel system fostered the design of the No. 1 Crossbar system. The demand for telephone service in the large metropolitan areas was so high, however, that No. 1 Crossbar production was used almost exclusively in new installations rather than in replacement. The first office was cut into service in Brooklyn, New York in 1938, and some 350 systems were installed during the next 10 years.

The two most significant features of No. 1 Crossbar in comparison to its predecessors are the network switch and the *marker* common control.

The network switch is classified as *fine motion* and makes use of a sequential, mechanical, logical "ANDing" at the coincidence of two orthogonal crossbars capable of very limited rotary motion (see Figure 3.9) to make a selection at each switch in defining a path through the network. Briefly, a horizontal selecting bar is rotated a few degrees either upward or downward, deflecting steel wire fingers at each crosspoint position. A vertical holding bar is then rotated, applying pressure to all the fingers in its column. Only the finger at the selected horizontal level will transmit the applied pressure to the crosspoint contacts, causing them to close. After the selection is made, the horizontal bar is released and is free to make other connections; the vertical bar remains energized (dissipating holding power) for the duration of the connection.

FIG. 3-9. A 10X20 Crossbar switch.

The upper half of Figure 3-10 depicts a crosspoint (consisting of three contacts) viewed from above and from the side, looking down the selecting [horizontal] bar in the normal (unoperated) condition, with the upper crosspoint (shown) operated, and with the lower crosspoint (not shown) operated. The lower portion of the figure is a perspective view of the crosspoint pair with the selecting finger in the normal (solid line) and rotated (dashed line) positions.

The flexibility of the selecting finger is an essential element in permitting the horizontal bar to be used in closing crosspoints at other intersections with other vertical members without disturbing the force-translating task of any of its fingers that are engaged in holding crosspoints closed. The damping spring serves the purpose of preventing excessively protracted mechanical oscillation of the selecting fingers when they are released to open a crosspoint.

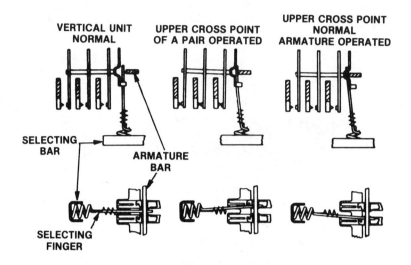

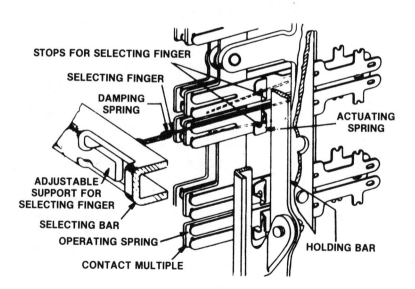

FIG. 3-10. The Crossbar switch mechanism.

The crosspoints employed in Crossbar switches were the first to use precious metal (palladium) contacts, which are electrically quieter. The corrosion resistance of the contacts, and the fact that they were *bifurcated*, have proven a great aid to reliable switching. A bay of crossbar switches is shown in Figure 3-11, illustrating continuation of the 11-foot standard originally set for SXS.

FIG. 3-11. A Crossbar switch bay.

The control consists of a collection of two types of markers or controllers (originating and terminating), both groups of which share the traffic load among their members on a direct basis (in the parlance of modern multiprocessor systems) but share it on a functional basis between groups. Each marker is capable of advancing the state of a call, but does not remain with a call for more than a fraction of a second. Each intraoffice call is served by two markers during its setup, one of each type.

The number of markers needed increases with the traffic in an office (the range is 3-to-10). Any marker detecting difficulty within itself generates a trouble record.

The philosophy of design of No. 1 Crossbar is similar to that of the Panel system in that it assumes that the majority of traffic handled will be interoffice.

Therefore it also splits the switching machine into two parts, originating and terminating, the latter communicating with its originating part and other offices on the same basis (see Figure 3-12). It differs, however, in that lines have a single appearance on one side of the network for both originating and terminating purposes, so that an intraoffice call path traverses the line link portion of the network twice (i.e., the network is folded with respect to intraoffice traffic).

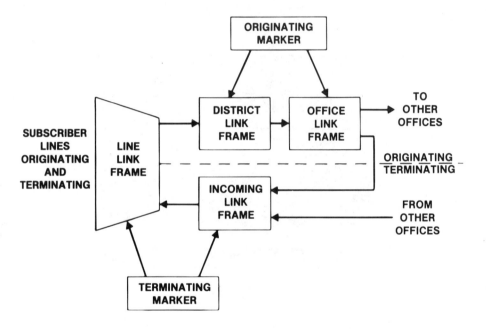

FIG. 3-12. No. 1 Crossbar organization.

Briefly, the originating portion is served by originating senders and markers. The sender accumulates dialed digits after returning dial tone, and after a marker is called in to set up the connection (entailing some six stages of switching) via an intraoffice trunk to the terminating portion, it "sends" the dialed digits to the terminating sender, which calls in a terminating marker to complete the connection (through four additional stages of switching).

Because of the strong Panel environment at the time of the design, the originating and terminating senders communicate via revertive pulsing,[3] which thus allows the originating portion to treat a distant Panel office for outgoing traffic in the same manner as it does the terminating portion of its own machine for intraoffice traffic; similarly, the terminating portion treats incoming traffic from a Panel office the same as it does intraoffice traffic. (Later versions provide MF pulsing capability for interoffice calls.)

After having performed their duties, the originating and terminating markers drop off; they are individually involved with a call for about one-half second each. Similarly, the originating and terminating senders disengage, the former after transmitting the last four dialed digits, the latter after signaling the incoming trunk circuit to ring the called line or to return busy to the calling line.

Unlike the Panel and SXS systems, the speech path is searched in parallel rather than sequentially, and a second trial can be made if the first attempt to set up a path fails, allowing more efficient use of network crosspoints. Many of these machines are still in service, but they are being rapidly supplanted by electronic systems.

No. 5 Crossbar

The success of No. 1 Crossbar led to the design of another system using similar philosophies but intended for a more attenuated application: No. 5 Crossbar. The No. 5 machine was originally to fulfill the needs of suburban applications, i.e., intermediate size offices with low and primarily intraoffice traffic. The first office was cut into service in 1948, in Media (a suburb of Philadelphia), Pennsylvania. Through the years, the No. 5 machine has been expanded into versions that span the range of application from the CDO of a few hundred lines, through large metropolitan offices of up to 40,000 lines, to class 3 and 4 toll offices.

Compared to No. 1 Crossbar, the markers of No. 5 are given more of the control responsibility and the sender functions are reduced and modified such

---

3. Recall that revertive pulsing entails the sending office's generation of a start pulse, and the receiving office replying with a train of pulses; when the sending office has counted the requisite number to be conveyed, it sends a stop pulse.

that *originating registers* accept dialed digits and need only transmit information to a sender when the call is outgoing, senders being used only for transmission of information to other offices; *incoming registers* accept signals from incoming trunks. Though a single (*combined*) marker type was originally used, later versions make use of two types, *dial tone* and *completing* with tasks implied by their names (see Figure 3-13). Registers and senders are shared facilities in the same fashion as senders in No. 1 offices.

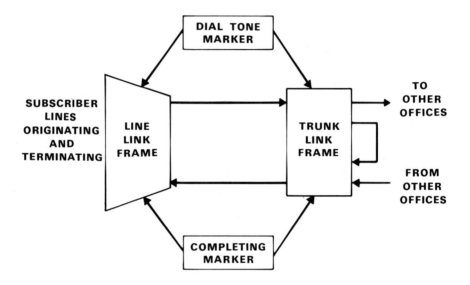

FIG. 3-13. No. 5 Crossbar organization.

The expectation of suburban application, with its attendant relatively high percentage of intraoffice traffic, fostered a design effort to make the handling of that component of traffic as efficient as possible. Thus the originating register retains the dialed directory number for an intraoffice call, rather than forwarding it to a separate register, saving the transmission time (in effect, the originating register plays both roles, first that of an originating sender, then that of a terminating sender).

There are over 2,000 No. 5 Crossbar machines in service at this writing, serving a large but diminishing percentage of the Bell System Lines.

No. 5A Crossbar

The 5A Crossbar system is based upon the No. 5 Crossbar, but with reduced feature offerings, and a miniaturized switch. The machine is offered in two sizes, 980 and 1960 lines, and can grow from the smaller to the larger size.

ESS

The advent of electronics offered the possibility, and of transistors the feasibility, of a switching system of profoundly different makeup and capability.

The electromechanical systems then in existence, though remarkably capable, tended to be inflexible in the sense that alterations in their functions required, at best, physical changes in the hardware, and some desirable new features were virtually impossible to provide.

Stored-program control (SPC) electronic switching offered far greater flexibility and maintainability, and the promise (ultimately) of greater economy and capacity than any other approach. The former two advantages are the result of the stored-program approach, the latter two are due to the solid-state electronic implementation (i.e., a wired logic electronic implementation is also possible[4] and would possess the potential for economy and high capacity, but it would remain as inflexible and difficult to maintain as its electromechanical forebears).

The major impact of electronics in switching to date has been in the control area, where the high-speed of the componentry has been exploited. Recall that both Panel and Crossbar systems make use of centralized control but require a multiplicity of control elements (senders or markers, respectively) to handle any reasonable amount of traffic because of the limited speed of the electromechanical devices (operation times of the order of milliseconds). With devices operating in nanoseconds, it became possible to design a single control capable of performing the same tasks (and many more) that required several controls in electromechanical systems.

Philosophies and specifics concerning stored-program control will be discussed in Chapter 6; the following will accordingly limit details of ESS machines to a level paralleling that provided for electromechanical machines.

---

4. And indeed has been applied in non-Bell systems.

No. 1 ESS

Following research dating back into the 1940s (see Figure 3-14), exploratory development of an electronic central office began in the mid-1950s, leading to a field trial in Morris, Illinois, in 1960. This trial proved the feasibility of the stored-program concept in a telephonic environment and fostered the development of No. 1 ESS.

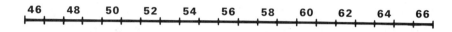

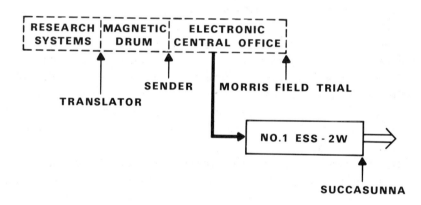

FIG. 3-14. No. 1 ESS genesis.

The Morris system was unique in its application of many techniques previously untried in the telephone sphere:

Stored-Program Control
Gas Tube, End-Marked Network
Flying Spot Store
Barrier Grid Store
Tone Ringing.

The choice of memory (store) technology was not inappropriate to the era of the inception of the system design (neither memory implementation survived into the development system; they are listed for completeness).

The network implementation likewise was a noble experiment that was supplanted. Certainly, its *end-marked* feature (such a network sets up a path autonomously in response to the electrical selection [marking] of two endpoints) was a worthy impetus to its adoption for an exploratory system. Its major disadvantage (aside from cost) was its inability to carry conventional ringing signals (this was the reason for employing tone ringing), which in turn required special station sets. The stored-program concept was the most robust of the new techniques tried in the Morris system.

Based upon the favorable results of the Morris trial, No. 1 ESS was developed, with its first office cutover (there was no field trial) in Succasunna, New Jersey, in 1965. The approach taken in the design of No. 1 ESS was to absorb into the control, responsibility for as much of the timing and decision-making activities as possible. This philosophy saved per-line and per-trunk hardware as well as shared-function hardware, at the expense of program memory and processing time; as the size of an office grew, this trade-off became increasingly attractive (until the processing capacity was exceeded). The system organization is outlined in Figure 3-15.

The network chosen employs sealed dry (as opposed to mercury wetted) reed switches (Figures 3-16 through 3-18), which are classified as *fine motion*, and which are activated or deactivated by brief current pulses (9 amperes, 300 microseconds) that change the remanent state of square-loop magnetic material associated with each pair (for 2-wire) or quad (for 4-wire) of reed capsules; the capsule, magnetic material, and winding assembly is called a *ferreed*. Since the switch state is defined by the remanent state of the associated material, no power is consumed in holding a switch open or closed. No *sleeve* lead is employed to permit direct physical path hunting; rather a

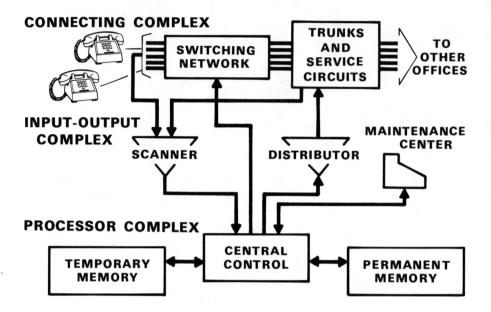

FIG. 3-15. No. 1 ESS organization.

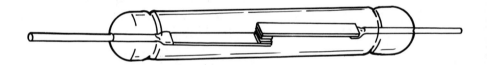

FIG. 3-16. The sealed reed switch.

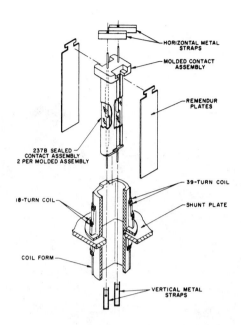

FIG. 3-17. Exploded view of ferreed.

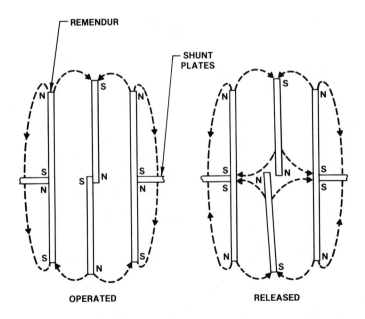

FIG. 3-18. Flux paths.

network map is retained in memory and kept up to date as changes occur. Figure 3-19 shows the basic 8-by-8 ferreed structure.

FIG. 3-19. An 8X8 ferreed switch.

The network utilizes eight stages of switching split into a 4-stage line link network that is *folded* (the speech path traverses the network twice, once in each direction) with respect to intraoffice traffic, and a 4-stage trunk link network that is folded for tandem traffic (Figure 3-20). Figures 3-21 and 3-22 display details of the line link network and trunk link network (the nature of these elements will be better understood after Chapter 5 is studied).

Nonremanent magnetic material is employed in a saturable transformer-like configuration (Figure 3-23) in the *ferrod* scan points, which replace the line relays of prior systems. A brief scanning pulse is applied to the primary winding of the ferrod, and an output pulse appears on the secondary winding if no line current is present to saturate the magnetic material.

The system is depicted in more detail in Figure 3-24 and can be seen to consist of a central control, a set of program stores, a set of call stores, the network, the scanner, and means for communication with the peripheral elements. Discrete component DTL (diode-transistor logic) was employed.

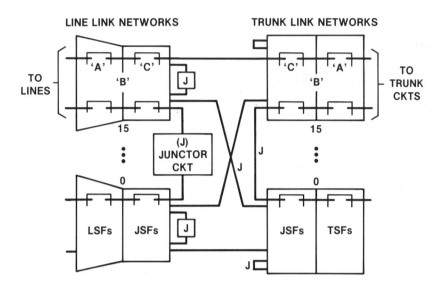

FIG. 3-20. No. 1 ESS network.

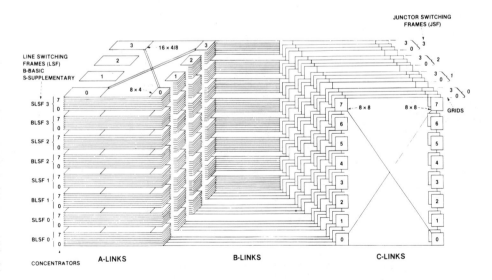

FIG. 3-21. No. 1 ESS line-link network.

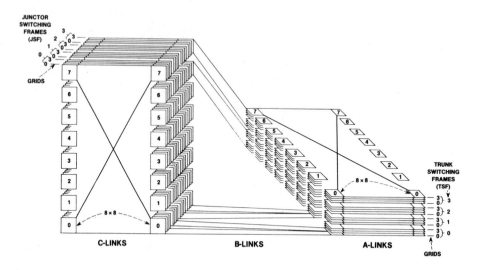

FIG. 3-22. No. 1 ESS trunk-link network.

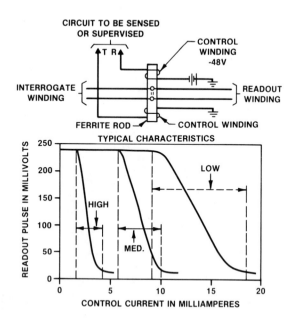

FIG. 3-23. The ferrod sensor.

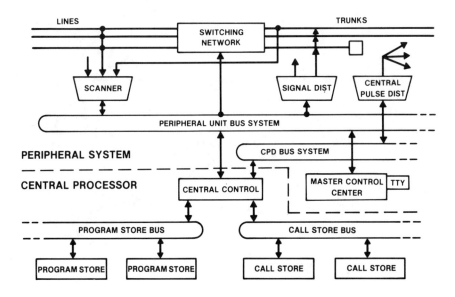

FIG. 3-24. No. 1 ESS block diagram.

The separation of program and call store is not necessary from a functional point of view, but it speeds processing and permits different technologies to be applied to optimize cost and reliability. The program store makes use of small permanent magnets and *twistor* sensing; the memory is not directly alterable electronically while in place, but can be changed by mechanically replacing the magnet cards which can be rewritten off-line on a special device. Its function is to retain the stored program and other information (e.g., *translations*) which remain relatively static.

The twistor principle was originally discovered in the study of the properties of a hard magnetic wire that had been subjected to torsional strain. It was found that the easy axis of magnetization was geometrically distorted into a helix, so that an incremental length of wire encompassing one turn would virtually constitute a magnetic torus and could be employed in a magnetic core-like manner. In practice, manufacturing difficulties led to use of a copper wire wrapped with a spiral of flattened hard magnetic wire (Figure 3-25).

Figure 3-26 illustrates the construction of the memory, which employs the twistor elements as sensors of the presence or absence of tiny permanent magnets that define the memory contents. When a word of memory is accessed

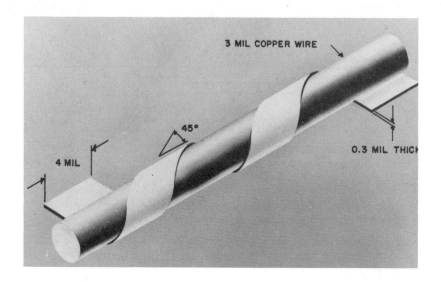

FIG. 3-25. The twistor element.

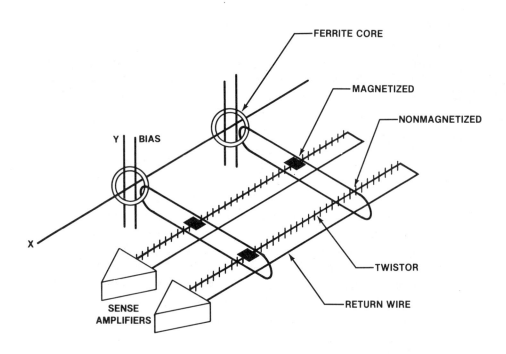

FIG. 3-26. The twistor memory.

via the correct X and Y drive pulses, whose combined magnetomotive force is sufficient to switch the large word-drive core, the resulting current in the selected solenoid tape is sufficient to switch the portions of the twistor wires adjacent to it from a reference remanent flux state to the opposite remanent flux state if they are not "paralyzed" by the presence of a coding magnet. (The coding magnets are "absent" in the sense that they are not magnetized). The coding magnets magnetically bias the twistor element so that it cannot be switched by the solenoid current.

The contents of the memory are altered mechanically by extraction of the coding cards (Figure 3-27), recoding and reinsertion.

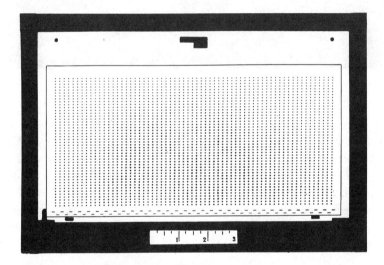

FIG. 3-27. The twistor card.

The memory is thus read-only (PROM) and nonvolatile under power failure. A typical office installation is shown in Figure 3-28. The store's capacity is 130,000 words of 44 bits each, and typical installations require 4 to 12 stores depending upon the program size and the number of lines and trunks. Only 37 of the bits are employed for information, the remainder being used for a double-error detecting, single-error correcting *Hamming code* check over the information, the check code bits, and the address. The cycle time of the memory is 5.5 microseconds (cycle time includes read and reset).

FIG. 3-28. The twistor store.

The call store, which holds dynamic information (e.g., the network map), must be electronically alterable, and makes use of ferrite apertured sheets which function in a manner analogous to magnetic cores. Figure 3-29 displays the ferrite sheet in three stages of manufacture—the pressed "green" ferrite sheet, the fired sheet (note shrinkage), and the completed, metallized sheet.

The philosophy of this device recognizes that a small annulus of material about each aperture will function like a magnetic core if drive currents are not excessive, and one of the windings necessary for memory operation can be brought into being via two-sided plating and plated-through holes. The memory module has a capacity of 8,192 words of 24 bits (23 bits + parity); its cycle time matches that of the program store. A less expensive and physically smaller ferrite-core memory replaced the ferrite sheet call store in later installations.

A brief outline of the processing of an intraoffice call follows (see Figure 3-30):

> An origination is detected during a scan of the line ferrods, and the change of state is noted in memory. A search is made for an idle

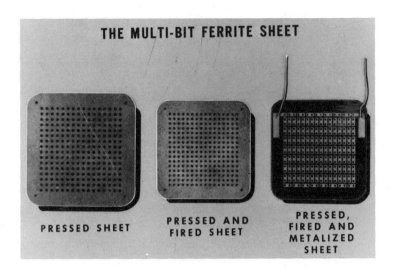

FIG. 3-29. The ferrite sheet.

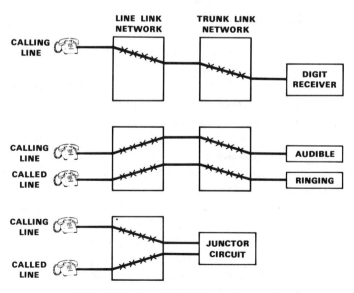

FIG. 3-30. No. 1 ESS intraoffice call.

receiver and for a network path from the originating line to the receiver, and the path is set up. After some tests, dial tone is supplied by the receiver, and the receiver is scanned for dialed information. When dialing is complete, the receiver is released, and an idle ringer and audible tone source are searched for, as are paths between the called line and the ringing circuit, and the calling line and the audible source, then the paths are set up. When the called party answers, these sources are released and a previously searched for and reserved path is set up between the two lines via a junctor circuit.

An interoffice call entails (after dialing) setting up a path through the trunk link network between a sender and a trunk to permit forwarding the address information to the distant office, then setting up a path from the calling line to the trunk.

The actions described above do not actually take place in a smooth, coherent fashion. They are the net effect of many distinct programs which (in general) perform the same segment of the total task for a number of calls being entertained in the system, but which may or may not advance the state of a given call, at a given time that they are run.

The use of a single control brought with it the possibility of total office failure hinging upon any of tens of thousands of single component failures. The nature of prior switching machines was such that total office failure was a truly rare event usually associated with a massive catastrophe such as a flood or earthquake. The early systems exhibited such robustness because of their distributed nature and redundancy. The performance of existing machines prompted the goal for ESS of 2 hours of office downtime in 40 years, an objective which has been closely approached.

Selective duplicative redundancy is employed in ESS to attain the necessary availability (other techniques are also possible, e.g., triplication with voting, but were rejected for this application). The duplicates run in synchronism, processing the same instruction simultaneously, and a match circuit monitors for disagreement. When a disagreement occurs, there is no immediate way of knowing which machine is at fault, and a specialized, nimble (because processing of calls must not be suspended for long) program is utilized to determine via gross checks, the relative sanity of the two controls. After a judgment is made, the sane control begins processing calls, and in its spare time performs diagnostic tests upon the suspect machine, generating data which (ideally) enables a craftsperson to resolve the problem to within a few replaceable circuit cards via a fault dictionary.

Other trouble detection techniques employed include:

Code checks (e.g., simple parity, Hamming),

Handshaking (acknowledgment by one portion of the system of receipt of information from another),

Timers (e.g., *watch-dog* timers which must be reset by the system periodically),

Sequencer checks,

Audits (an extremely important feature which combs through call store making "reasonableness" checks and resetting unjustifiable conditions).

An elaborate hierarchy of alternative procedures exists, available in proportion to the magnitude of the difficulty the control may find itself in. These range from switching subsystems about in search of a viable arrangement, to a cold restart, relinquishing all confidence in the system's state indicators.

In large No. 1 ESS offices, an additional (duplicated) processor known as the Signal Processor (SP) is employed to relieve the No. 1 control of the peripheral access and control tasks. SP-equipped offices have about twice the capacity of nonequipped offices.

Refinements in programming over the years—in spite of many feature additions—also have about doubled the capacity of No. 1 ESS so that, for example, an office equipped with an SP today has about four times the capacity that an office without an SP had in 1966.

There were 885 No. 1 ESS machines in service in May, 1982.

No. 1A ESS

The 1A Processor was developed in response to the needs of both local and toll applications (the latter will be discussed in Section 3.2). The local, 2-wire machine utilizing this processor is called No. 1A ESS, and was first cutover in Chicago in 1976.

There was a need for machines with greater line capacity than No. 1 ESS could accommodate even with an SP. In the period from 1960 to 1970, the number of Bell System lines increased by about 70 percent, while the number of buildings increased by only about 10 percent. In fact, about half of the growth that took place in the 1960s was absorbed in about 6 percent of the buildings. These buildings grew larger primarily by increasing the number of entities (switching machines) on the premises, and multiple entity buildings are now quite common, especially in metropolitan areas.

Multiple entities are disadvantageous because they cause *trunk splintering;* must provide interentity trunks; require multiple work forces, traffic service centers, and engineering effort; and fragment new service offerings during transitions. (Trunk splintering is the effect of having multiple, small, inefficient trunk groups from a distant office *homing* on multiple entities in a building rather than a single, large, efficient trunk group homing on a single entity.) The processor was also used in retrofitting capacity-limited No. 1 ESS offices (the first retrofit took place in 1978).

The 1A machine uses T^2L logic and originally employed ferrite-core memories for both program and call stores (semiconductor memory later became standard).

Though there was some original interest in making No. 1 and No. 1A totally program-compatible, it was decided that such a requirement would unduly restrict the structure of the 1A, and upward compatibility was deemed adequate. Further, it was decided that bit-to-bit upward compatibility was unnecessary and some program transformation via utility programs was acceptable.

Thus, though the two machines are quite distinct, upward compatibility was attained. On the other hand, maintenance programs for the processor itself cannot be made compatible because the machines' implementations are different. Nonetheless, the philosophies of these programs can be carried over, and maintenance programs for the rest of the system are not materially affected.

The major reason for the difference in performance between the two machines is the device and circuit technology, a change from discrete diode-transistor logic (DTL) to integrated transistor-transistor logic (T^2L), and from twistor memory to 2-1/2 D core. The higher-speed logic (7-nanosecond gate delay as opposed to 40-nanosecond) and memory (1.4-microsecond cycle time as opposed to 5.5 microsecond) and reduced size (and therefore propagation delay) combine to yield a very-high-performance machine.

The remarkable thing to note about devices and circuitry over the past two decades is that better equipment (i.e., faster, smaller, dissipating less power) is not more expensive, but is in fact actually cheaper (inflation not withstanding) when it becomes available. Telephone equipment, as well as other electronic-based products, is a direct beneficiary of this phenomenon.

Other features of the processor include multiple general registers, subroutine return address pushdown stack (also in No. 2 ESS), single- and double-word instructions, *shadow* registers, which save active register contents at the programmer's behest, and reduced high-speed memory requirements because of the writable program store with disk and tape backup.

The network employs *remreed* crosspoints, which utilize hard magnetic reed blades rather than hard magnetic material around the crosspoints, increasing the sensitivity of the switch and reducing its size. Remreed frames are being installed in No. 1 ESS offices as well.

There were 750 1A ESS systems installed by May, 1982.

No. 2 ESS

Figure 3-31 traces the genesis of the No. 2 ESS machine, beginning with research systems leading to the No. 101 ESS, a stored program PBX,[5] and eventually the subject machine.

No. 2 ESS is a machine intended for medium size ($\simeq$ 1,000 to 10,000 lines) suburban office application (as was No. 5 Crossbar). It differs from No. 1 ESS in size, control architecture, network arrangement, machine language, and program philosophy.

It shares with No. 1 ESS the dual-memory (program and call store) approach (twistor and ferrite sheet, respectively), ferreed network elements, ferrod scanning elements, and many other details. Hybrid Resistor-Transistor Logic (RTL) was employed because it was less expensive than DTL and fast enough for the application.

The control uses one address per instruction as does No. 1 ESS, but unlike No. 1, two instruction lengths are employed, full word and half word. This feature is believed to have been instrumental in reducing the amount of program memory required (relative to No. 1). It also employs a *push-down*

---

5. Private Branch Exchange.

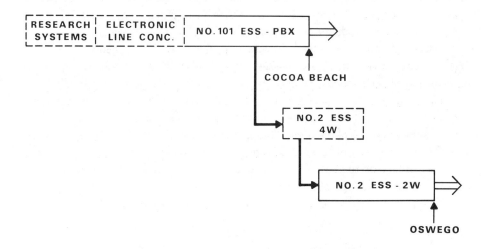

FIG. 3-31. No. 2 ESS genesis.

stack[6] mechanism for subroutine return addresses, and a *transfer allowed* bit in each instruction to indicate whether the instruction is a legitimate destination of a transfer.

There are no index registers, though means are provided for replacing the low-order portion of an address with a selected field rather than adding arithmetically.

The network employs four stages, folded to produce eight stages of switching for lines and trunks (see Figure 3-32).

There were 235 No. 2 ESS offices in service in May, 1982.

No. 2A ESS

A modularized form of No. 2 ESS called No. 2A ESS was made available for rapid shipment and setup.

---

6. A last-in first-out storage mechanism analogous to a cafeteria tray dispenser.

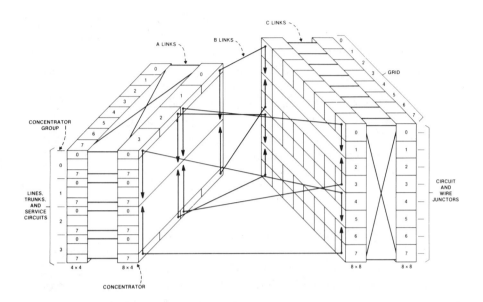

FIG. 3-32. No. 2 ESS network.

No. 3 ESS

Until the 1970s a portion of the switching system market that had remained untouched by electronics was the rural Community Dial Office (CDO). This is the small (100 to 1000+ line) office typically situated in a small town in an agricultural area. It is characterized by low calling rates and chiefly intraoffice traffic. A small percentage of Bell System lines are terminated on CDOs, while about half of Bell System buildings are CDOs.

Historically, SXS had totally dominated this market because a distributed control system has a very low "getting started" or 0-line cost, and SXS grows very gracefully in small offices. In contrast, a common control system has the expense of its control ballooning its first cost, and though its incremental cost per line may be lower than SXS, the initial upward bias in cost tends to price it out of the market for the entire small office size range:

Work began on an Electronic Community Dial Office (ECDO) using electronics in 1966, and took on a form intermediate between wired logic and stored program. This evolved into a stored-program machine that has been given the name of No. 3 ESS. It offers ANI, TOUCH-TONE, alternate routing, AMA, and remote maintenance and administration.

The No. 3 ESS Processor (3A CC) is microprogrammed, and uses integrated circuit logic. The microprogram Read-Only Memory (ROM) employs bipolar integrated circuits, while main memory uses IGFET ICs. Both program and call store reside in the same memory, and a tape cassette backup is provided.

The No. 3 ESS system proper has a remreed network and ferrod scan points. It is shop-assembled and tested and shipped in modules that are connectorized for rapid installation.

The first No. 3 ESS was cut over in Springfield, Nebraska, on July 31, 1976, and 275 systems were installed by May, 1982.

No. 2B ESS

The No. 3 ESS Processor, as has been mentioned, is a microprogrammed machine capable of *emulating* other machines. That is, its microprogram memory can be reprogrammed so that it can accept the binary machine language of another machine and perform the same net functions (intermediate functions may differ). (In the same fashion, for example, the IBM/360 Model 65 was capable of emulating the IBM 7094, though the machines' inner structures differed dramatically.)

Since the No. 3 Processor has greater throughput capability than the No. 2 Processor and is less expensive, it is reasonable to consider using the No. 3 machine to emulate it, profiting from the greater call-handling capability and economy without any reprogramming of the call-handling software.

The system which results from employing the No. 2 ESS periphery and programs with a No. 3 Processor is called the No. 2B ESS. The first system was cut over in Acworth, Georgia in early 1976, and 470 systems were in service by May, 1982.

No. 5 ESS

No. 5 ESS, the newest of the Bell line of switching products, is best understood when placed in perspective relative to its technical anticedent, No. 4 ESS. No. 5 is therefore discussed in Section 3.2.

3.2 TOLL

Through the 1930s, toll traffic was handled almost exclusively by manual operators (some SXS was so employed). During the depression era, however,

BTL began limited No. 1 Crossbar (local) development, and in later 30s, Crossbar Tandem and No. 4 Crossbar development began (see Figure 3-33).

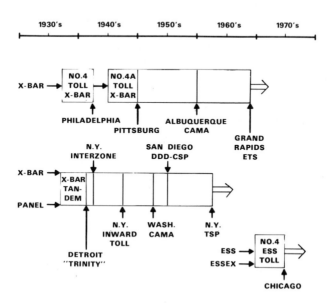

FIG. 3-33. Toll machine genesis.

Toll offices switch traffic between areas often separated by substantial distances, and may have trunks to tandem offices and nearby local offices as well as other toll offices (see Figure 3-34). Toll offices generally provide 4-wire switching for better long-haul transmission.

Only six No. 4 offices were installed in the early 40s, and after World War II, a modification was made to aid nationwide operator distance dialing and allow customer *Direct Distance Dialing* (DDD); the modified system was renamed the No. 4A Toll Crossbar System. The six No. 4 offices were retrofitted with DDD features and renamed No. 4M. Over the years, Crossbar Tandem systems have been modified and expanded so that their capabilities heavily overlap those of the 4A machines, but their primary function remains the switching of 2-wire interlocal office traffic. Among the key capabilities of 4A machines are full *6-digit translation* (on 10-digit numbers) and *expanded alternate routing, variable spilling,* and *code conversion* features.

Six-digit translation, made possible initially by the advent of the *card translator* (the first Bell System application of the transistor), allows the

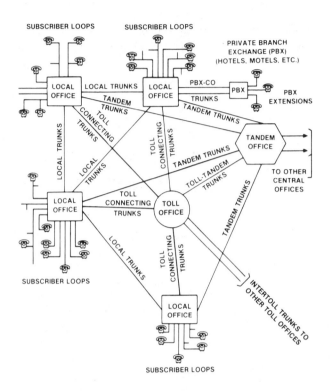

FIG. 3-34. Typical trunk patterns.

machine to choose more precisely the best route to be taken, because the specific office with the destination *Numbering Plan Area* (NPA) can be identified. (The card translator was primarily electromechanical.) The Electronic Translator System (ETS), using the SPC 1A stored-program processor, was introduced in 1969, and replaced virtually all the card translators in Bell System Crossbar Toll offices by 1977 (a number of 4A machines are owned by non-Bell companies).

Alternate routing permits an orderly progression of possible routes to be tried when all busy trunk groups are encountered. Generally, the order is from the most direct to the most circuitous, progressing up the hierarchy in the class of switching machines involved. (The alternate-routing discipline will be described in some detail in Chapter 4.)

Variable spilling pertains to the ability of the system to absorb three or six digits and spill the remaining digits forward. Code conversion is the substitution or addition of leading digits before spilling. Absorption of digits is performed when the geographic area of the office reached is that identified by the leading digits, which can then be safely deleted. Substitution of digits is performed when the nominal route is blocked and an alternate route is chosen, and leading digits are added when it becomes necessary to route the call out of an area, then back into it due to blocked direct routes.

No. 4 ESS

Toll traffic had been increasing at a high rate and was expected to continue to do so, requiring a continuous and massive increase in the amount of equipment in the DDD network. The primary machine available for this application was for many years the 4A Crossbar System which, because of its finite (though large) capacity, was becoming unduly numerous, complicating administration and reducing trunking efficiency (large trunk groups are more efficient than small ones). A new machine with a very much larger capacity was needed; hence, the No. 4 ESS.

Figure 3-35 illustrates in oversimplified form the trunking advantages of a large machine.

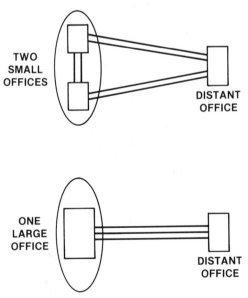

FIG. 3-35. Large toll office advantages.

If a toll switcher has exhausted, necessitating a second office, there must be separate trunk groups (splintered trunks) to distant toll offices, and trunks between the old and new offices. If a single large toll office is employed instead, a single trunk group with fewer total trunks than the aggregate of the splintered groups it replaces will suffice, and maintenance and administration is reduced.

The time-division network of No. 4 ESS, a radical departure from past systems, was chosen because of the very large number of terminals to be handled (over 100,000), and the increasing percentage of digital trunks. It employs two varieties of time-division switch implementation (see Figure 3-36), one the time-shared use of a space-divided network (the center stages), the other a *time-slot interchanger* (the first and last stages).

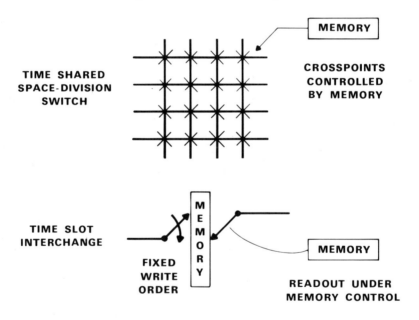

FIG. 3-36. Time-division switch elements.

The nature of such a network may be best understood by first considering the simple "network" of Figure 3-37. Here we see analog inlets being sampled in strict rotation, with the sampled amplitude converted to digital form and stored in corresponding successive locations in a memory. At the outlet side, digital samples are extracted from memory, converted to analog, and distributed in

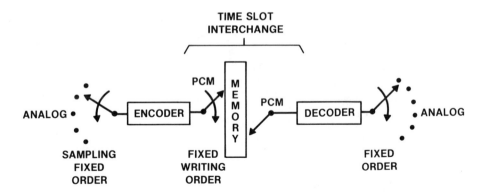

FIG. 3-37. A simple PCM network.

strict order to the outlets. It will be noted, however, that the order of extraction from the memory will not be sequential, but rather will depend upon which inlet is to communicate (on a one-way basis) with which outlet. A control is therefore employed which correlates the sequential outlet number with the inlet that is to be connected and induces a memory read from the location containing the latest sample.

This arrangement might work very well, but the bandwidth requirements limit the number of terminals that can be handled. For example, the sampling rate must be at least 8000 Hz, so that a 0.5-microsecond memory could handle only 125 inlets, with half the memory cycles being *write* and the other half *read.*

Thus, if a number of modules of this kind are to be used to serve a large number of terminals, means must be provided for intercommunication among them so that an inlet connected to one module may communicate with an outlet of another module. A *time-multiplexed, space-division switch* serves this purpose admirably (Figure 3-38). (This TST [time-space-time] arrangement is not the only one possible; another major category is represented by the STS arrangement, as will be seen in Chapter 7.)

The need for buffering the PCM sampled information should be understood. Consider the case of a 24-channel PCM trunk between Kansas City and Chicago, and another one between Chicago and Columbus, Ohio. Suppose that someone in Kansas City whose call has been assigned time-slot 3 on the trunk

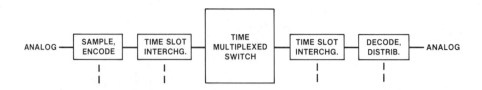

FIG. 3-38. A large PCM network.

to Chicago wishes to call someone in Columbus, but the trunk to that city has only time-slot 18 available. The call could not be completed unless a means existed to buffer (store) each sample coming in on time-slot 3 from Kansas City until time-slot 18 arrived, so that the sample could be shipped to Columbus. The memories in the Time-Slot Interchanger perform that function.

The network for the initial application (the Chicago 7 office, cut over in January, 1976) at first interfaced only with analog inputs; digital trunks underwent demultiplexing and decoding to analog before entering the network. The network itself was therefore of the form shown in Figure 3-39.

It will be seen that there are 128 groups, each consisting of seven subgroups of 120 analog terminations each, for a total of 107,520 terminations. The analog signals are sampled, converted to PCM, multiplexed (interleaved in time), and appear on busses accommodating 128 time slots (the additional eight slots are used for maintenance and similar purposes). These busses enter a distributor or *decorrelator*, which protects the network from highly correlated traffic, buffers the incoming data, and deloads the busses.

These three functions are performed using two groups of memory modules, the incoming group consisting of seven modules, the second group of eight modules. Each module is a 128-word by 9-bit memory. The time-slot information is deposited sequentially in the first module group and distributed, in a fixed manner, among locations in the second group. The accessing channels for the second group are thus deloaded because each module has only seven-eighths as many samples as the first group (105 versus 120).

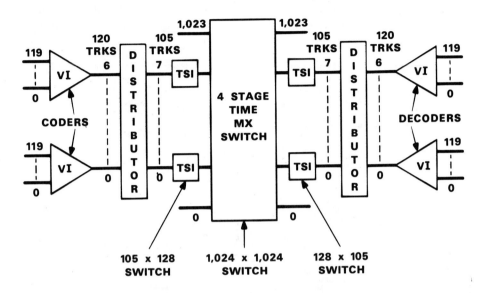

FIG. 3-39. No. 4 ESS functional view.

The *Time-Slot Interchange* (TSI) reads out of the second module groups in response to an ordering dictated by the control in establishing the required time-space path from inlet to outlet.

The *Time-Multiplexed* (space-division) *Switch* (TMS) center portion consists of four stages, the first and last made up of 8-by-8 grids physically associated with the TSI frames, the second and third made up of 256-by-256 grids built in turn of 16-by-16 switches. The entire switch is reconfigured every 976 nanoseconds. The 4-stage network provides 16 possible paths to each time slot, so that with 128 possible time slots for each path, there are 2048 possible paths through the network for each inlet-outlet pair. The first-trial blocking with 0.9-erlang occupancy is 0.005, making it virtually nonblocking. Since the network is inherently unidirectional, two separate paths must be set up for bidirectional (4-wire) operation.

The TMS and TSIs are fully duplicated for reliability.

After the initial application, digital trunks were switched directly. Figure 3-40 traces four channels through a slightly simplified representation of the network.

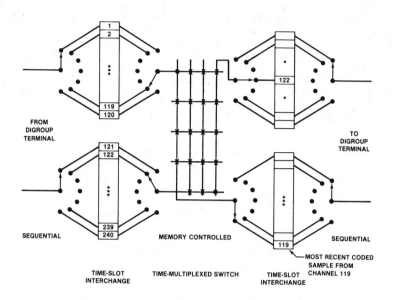

FIG. 3-40. Simplified TST path set up.

The 1A Processor is employed in the central control of this system, aided by signal processors in the periphery.

Almost 100 No. 4 ESS systems were installed by 1983.

No. 5 ESS

The advantages of digital switching that had been realized in No. 4 ESS were early recognized to have potential realization as well in the local switching arena. Significant problems existed, however, in local application.

Unlike the toll situation, where a large and growing proportion of the intraoffice trunks are already digital, no local lines are presently digital, and the deployment of such lines has not even begun. It is therefore necessary to perform the analog-to-digital and digital-to-analog conversion for each line. Compounding the problem is the fact that less capital can be reasonably invested in per-line circuitry than in per-trunk circuitry.

Similarly, because digital crosspoints are ineffective in carrying conventional ringing current, a separate ringing-access network becomes necessary.

Advances in semiconductor electronics offered answers to the dilemma of local digital application. Higher levels of integration made A/D and D/A converters less expensive, and new techniques produced semiconductor devices capable of handling ringing potentials and current as well as the over-voltages allowed by conventional protectors. Inexpensive microprocessors made it possible to conceive of distributed control associated with portions of the network function, allowing a growth capability analogous to that of the SXS system.

The advent of low-loss, high-bandwidth fiber and fiber-optics devices made it possible to design remoted modules that could function as though they were in the central office.

Figure 3-41 depicts the system makeup of No. 5 ESS. The network is a TST configuration, as was true of No. 4 ESS, but with the TSIs associated with interface modules that are under the control of individual microprocessors. This allows a large portion of the control to be purchased in proportion to the network or number of customers, conferring a growth capability comparable to that of SXS.

A central processor facility is provided, but the bulk of its activities are associated with administration.

The TMS must grow with the office size, but it constitutes a relatively small part of the network.

Fiber optic links are used for conveying voice-path signals as well as timing and control messages. These links can be extended to remote units that may be very distant.

The first No. 5 ESS machine was cut over in Seneca, Illinois in early 1982.

3.3 OPERATOR

From the beginning of activity in automated switching, the primary goal was the reduction of the operator force. It is noted from time to time that if only manual operation were available, a large fraction of the U.S. population would have to be employed as switchboard operators.

The actual number of operators employed by the Bell System has risen at an alarming rate over the years in spite of automatic equipment (see Figure 3-42), and the cost of maintaining such a workforce is substantial. Recognizing

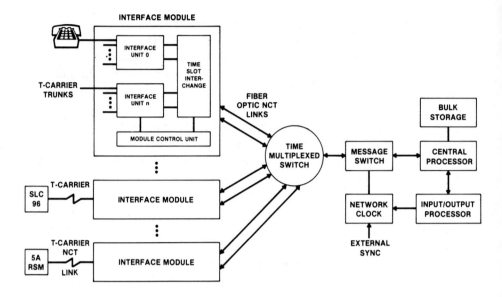

FIG. 3-41. No. 5 ESS system.

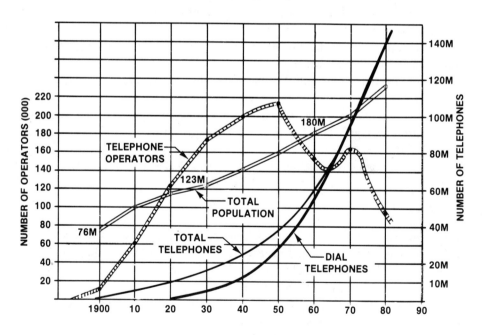

FIG. 3-42. Operator census history.

that some operator services are indispensable, efforts have concentrated of late upon increasing the operator's efficiency by providing equipment aids. Operator services include toll, coin, general assistance, directory assistance, and intercept.

Manual Boards

The manual cord switchboard has changed remarkably little in form or function over the years (see Figures 3-43 through 3-47).

Figure 3-43 depicts the Gold and Stock switchboard of 1879, where young men can be seen functioning as operators. Figure 3-44 shows the pyramid-shaped *Law Board* in Richmond, Virginia in 1882 (this was part of a system of communication between law firms). Figure 3-45 is reasonably typical of the many small, rural, family-operated telephone company switchboards of the early days. Figure 3-46 illustrates that both men and women were occasionally employed together in central office activities. Figure 3-47 indicates the interest in efficiency in switchboard operations (note the roller skates).

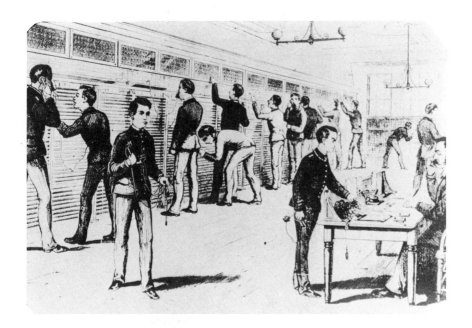

FIG. 3-43. Early central office.

FIG. 3-44. The "Law Board."

FIG. 3-45. An early rural central office.

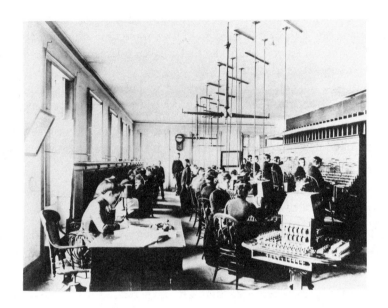

FIG. 3-46. A "coed" central office.

FIG. 3-47. A large PBX.

While the Bell System boards stopped at a density of 10,000 jacks per operator, some non-Bell boards have more than three times that number. Jacks are *multipled* across several boards so that a given origination can be served by any of several operators at the boards.

All non-DDD toll traffic must be handled by operators, many of whom still serve at cord switchboards. The tasks performed by these operators include:

1. Noting an origination, inserting a cord plug into the originating jack, and obtaining and recording the called and calling numbers (recorded by pencil upon mark-sensing cards).

2. Determining the *rate-and-route* information (often by calling a rate-and-route facility).

3. Recording the rate, inserting a cord plug into the appropriate trunk jack, keypulsing the route plus the called number, and marking the time at which conversation begins.

4. Noting the completion of the call, taking down the connection, and marking the time.

Automatic Call Distributors

Certain of the operator services such as directory assistance and intercept can be most efficiently handled if calls are automatically distributed among operators in a pool. Ideally, during high-traffic periods, an operator who had just finished serving one customer would be automatically connected to another.

No. 5 Crossbar introduced *Automatic Call Distributor* (ACD) capability in New York in 1968, and the feature is now widespread in metropolitan areas. In addition to the efficiency advantages afforded, this feature permits operator pools to be placed in convenient (e.g., suburban) locales. It is now offered as a feature on No. 1 ESS.

TSPS

The Traffic Service Position System (TSPS)[7] is an autonomous electronic

---

7. This system was preceded by the TSP system that is associated with Crossbar Tandem systems and is electromechanical.

switching system with a stored-program control, that provides highly efficient employment of operators for assisting customer-dialed toll calls. The controlling processor is the Stored-Program Control (SPC) No. 1A.

Briefly, the system has the capability of recognizing a dialed request for assistance (a dialed 0 preceding the 7- or 10-digit number), connecting an appropriate receiver to field the called number and the calling number (if the originating office has ANI[8] capability), and bridging a talking connection to an operator console (see Figure 3-48). The operator is automatically supplied with information such as the number of digits the customer dialed, and whether the calling line number must be obtained from the customer.

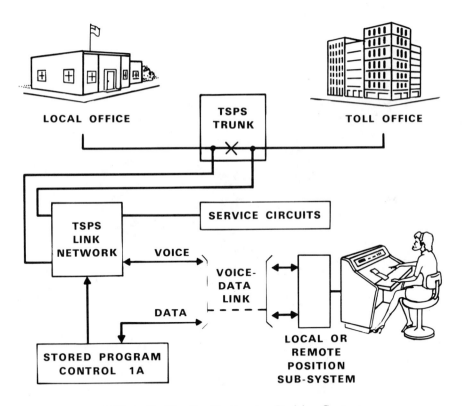

FIG. 3-48. The Traffic Service Position System.

---

8. Automatic Number Identification.

In a typical case, such as a person-to-person call, the system will cause the call to continue to progress through the network in parallel with the operator's oral communication with the calling party, so that the called party will be rung as the operator obtains the necessary information. Operator options include releasing the position from the call after performing the desired function, or remaining associated with the call (e.g., for time and charges), the system prompting when termination is detected.

Among the advantages of this system are the capability of placing the operator positions remote from the switching equipment, and the "human engineering" of the positions and their environment make the operator's job a relatively pleasant one (see Figure 3-49).

FIG. 3-49. TSPS consoles.

The effective range over which TSPS capability can be made available is extended greatly by the RTA (Remote Trunk Arrangement) system which is essentially a satellite concentrator.

Automatic Intercept System

It is necessary to provide a special information service called *intercept* for handling calls to numbers that have changed, been disconnected, etc. Calls to such numbers are automatically routed to an operator position where the necessary information is available from short-lived directories. The number of operators required is substantial.

With the goal of near total automation of this process, the Automatic Intercept System (AIS) was designed. It employs stored-program control (the No. 2 ESS Processor), a time-division network, and a voice-response unit, enabling automatic operation. When a call is routed to the system, the pertinent information about the called number is retrieved and relayed to the customer via sequential selection of the appropriate words and phrases from the voice-response unit. By remaining on the line, the customer gains access to an operator for further assistance.

The first system was installed in Hempstead, New York in 1970.

EXERCISES

1. The present numbering system used in all telephone systems is decimal. (There would be some advantages to having other systems, e.g., base 8 or 16; what might they be?) Suppose that a SXS switch costs $90 + $0.25b^2$, where b is the number base. (The $90 is for the frame and electromechanical moving parts, and the switch points cost 25 cents each; since the switch has b levels and b contacts on a level, that portion of the cost due to the contacts is $0.25b^2$.) Find the optimum b. Hint: Assume that you need to be able to store up to some large number N in a bank of SXS switches, each of which can account for two digits. In view of the cost/switch, what base would be optimum?

2. If an electronic SXS switch became available that performed identically to a conventional one at half the price, half the size, ten times the reliability and near zero maintenance, would it have a future in telephony? Why? Would it sell?

3. Is the number of Hamming check bits used in the No. 1 ESS program store exactly adequate for double error detection and single error correction over the information, check, *and* address bits?

4. Of the switching systems discussed, which is capable of most rapidly (after dialing is completed) ringing the called number for an intraoffice call? (Your intuition may fail you.)

5. Note that No. 4 and No. 5 ESS utilize two Time-Slot Interchangers. What would be the effect of reducing the number of TSIs to one?

6. In an interoffice call, which office returns the busy tone? Which returns audible?

7. What impediments are there to the introduction of optical fiber in the loop plant? Why are these impediments of little significance for interoffice trunks?

READING LIST

General

History of Engineering and Science in the Bell System — Switching Technology (1925—1975), A. E. Joel, Jr., Bell Telephone Laboratories, 1982.

Electromechanical Systems

Basic Telephone Switching Systems, 2nd ed. D. Talley, Hayden, Inc., 1979, Chapters 7, 8, and 9.

Switching Systems, AT&T, 1961, Chapters 5 and 6.

R. M. Morris, "Crossbar Tandem TSP," *Bell Labs Record,* May, 1964.

Electronic Systems

"No. 1 Electronic Switching System," *BSTJ,* Vol. 43, September, 1964, Special Issue in two parts; also issued as *Monograph 4853.*

A. H. Doblmaier and S. M. Neville, "The No. 1 ESS Signal Processor," *Bell Labs Record,* April, 1969.

"No. 2 ESS," *BSTJ,* Vol. 48, No. 8, October, 1969, Special Issue.

"The 1A Processor," *BSTJ,* Vol. 56 No. 2, February, 1977, Special Issue.

Basic Electronic Switching for Telephone Systems, D. Talley, Hayden, Inc., 1975.

"TSPS No. 1," *BSTJ,* Vol. 49, No. 10, pp. 2417-2732, December, 1970, Special Issue.

W. H. C. Higgins, "A Decade of ESS," *Bell Labs Record,* Vol. 48, No. 11, pp. 318-325, December, 1970.

G. D. Johnson, "No. 4 ESS—Long Distance Switching for the Future," *Bell Labs Record,* September, 1973.

H. E. Vaughan, "An Introduction to No. 4 ESS," *IEEE International*

Switching System Symposium Record, June, 1972.

A. E. Ritchie, "Evolution of Toll-Economics and Concepts," *ICC '77 Conference Record,* p. 179.

A Time for Innovation, A. A. Collins and R. D. Peterson, Merle Collins Foundation, 1973, pp. 32-39. (An analytical treatment of No. 4 and No. 5 ESS-type networks.)

F. T. Andrews and W. B. Smith, "No. 5 ESS - Overview," *ISS 81*, Vol. 3, pp. 1-6.

CHAPTER 4

CURRENT FEATURES
AND
APPLICATIONS

Switching system features include services provided to customers as well as functions performed in aid of the system itself, which are generally invisible to the customer. Applications are the employment of features.

4.1 MAINTENANCE

Maintenance for current systems falls into two categories: preventive and "upon demand." The first applies primarily to equipment (mechanical or electromechanical) with a wear-out mechanism which, to be thwarted, requires some recurring act, such as lubrication. The second applies to all systems since all fail occasionally, though the urgency of the activity may vary widely between systems. To prevent undetected failures from accumulating, equipment is periodically *routined* (caused to demonstrate its ability to perform).

In practice, many offices handling an appreciable number of lines are manned by maintenance personnel on a 24-hour basis. Minor failures may cause generation of a *trouble ticket* or similar indications that are routinely checked. When major failures occur, monitoring equipment rings one or more alarm bells and illuminates trouble lights to gain the maintenance personnel's attention. Unmanned offices activate alarms in a manned office. In electromechanical equipment, most failures reside in a sufficiently small or sufficiently redundant portion of the equipment that only a few or none of the customers are denied service or even inconvenienced by the malfunction; the diagnosis and repair can therefore proceed at a leisurely rate. In ESS offices, if a processor malfunctions, its twin will assume the load, but the pace of the repair activity takes on some urgency. The probability of a failure taking place in the healthy processor in the ensuing hour or so differs uncomfortably from zero, and that processor is the major diagnostic tool for finding the original fault. If both processors succumb, not only is the office incapable of serving

89

customers (a nearly unthinkable situation), but the prognosis for speedy revival is grave because a major tool (a sane, operating processor) to this end is lacking. (Imagine an insane psychiatrist attempting to analyze himself or, worse yet, two insane psychiatrists trying to analyze each other.) In practice, this has been a rare event, and tools and procedures are available for reviving even a dead processor pair.

The philosophy behind the abhorrence of an office becoming inoperative (one not always shared elsewhere in the world) is not merely due to the customer dissatisfaction (some of which is inevitable and regrettable, but tolerable), but due to the possibility of an emergency occurring in an area denied telephone service. If a customer's telephone fails him in an emergency and he can run next door or down the block and find a working phone, a potential tragedy might be averted; if the entire community is without service, the situation could be desperate. For ESS systems, therefore, the goal of an aggregate of 2 hours of total office downtime in 40 years was established as consistent with the grade of service to which customers had become accustomed.

To some extent this philosophy is the result of relatively ancient historical precedent. Manual systems were quite reliable; so long as office battery did not fail and one or more operators were present, service was provided. Step-by-Step equipment, though hardly failure free, is so distributed in nature that a malfunctioning switch disturbs service to only a few customers.

A corollary to this philosophy can be seen in electronic PBX[1] design where, in the interests of economy, a backup duplicate control may not be provided because phone service to the outside world would almost surely be available somewhere on the premises (in large PBXs, however, the rationale may revert to duplication). (In practice, some PBX-CO lines can usually be cut through manually in case of failure.)

The centralization of maintenance operations for a multiplicity of offices, including remote monitoring equipment and a staff of maintenance people who can be dispatched in case of troubles, is becoming more common.

---

1. Private Branch Exchange.

4.2 HIERARCHY

There exists a hierarchy of switching machines in the Bell System as alluded to in Chapter 2. It consists of five levels:

Class 5:   *End Office,* serves customer lines.

Class 4:   Acts as first stage of concentration for intertoll traffic from class 5 offices. There are two varieties, *Toll Centers* and *Toll Points*, which differ in operator services provided.

Class 3:   *Primary Center* (PC).

Class 2:   *Sectional Center* (SC).

Class 1:   *Regional Center* (RC).

The latter three office classes are known as *Control Switching Points* (CSPs) in the distance dialing network; these are switching centers at which intertoll trunks are switched.

The nominal hierarchical switching plan (Figure 4-1), though observed in the majority of cases, is not inviolable; any lower class office may *home* upon any office of higher class. (Every office of class lower than 1 in the hierarchy has a *final* trunk group to a single office higher in the hierarchy, and is said to *home* upon that office; this group is the last that may be tried if all appropriate *high usage* groups are busy.)

Regional centers serve geographical areas known as regions, of which there are ten in the United States and two in Canada (see Figure 4-2). The regions are subdivided into sections served by Sectional Centers that are subdivided, in turn, into primary areas served by Primary Centers.

The switching plan routing discipline is indicated in Figure 4-3, which illustrates the *final group* routing as well as possible *high-usage group* connections. High-usage trunk groups may be established between any two offices, regardless of location or rank, whenever the traffic volume justifies them; they carry the majority of offered traffic. When all trunks in a first-choice high-usage group are busy, the appropriate second-choice group is next tested, etc., until the final group is reached. Final groups are engineered for very low blocking probability, so that an attempt to use one is almost always successful.

There are some limitations to the extent that high-usage trunks groups may

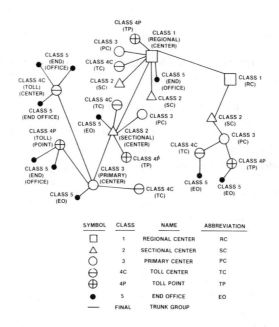

SYMBOL	CLASS	NAME	ABBREVIATION
☐	1	REGIONAL CENTER	RC
△	2	SECTIONAL CENTER	SC
○	3	PRIMARY CENTER	PC
⊖	4C	TOLL CENTER	TC
⊕	4P	TOLL POINT	TP
●	5	END OFFICE	EO
—	FINAL	TRUNK GROUP	

FIG. 4-1. Switching plan (basic principle).

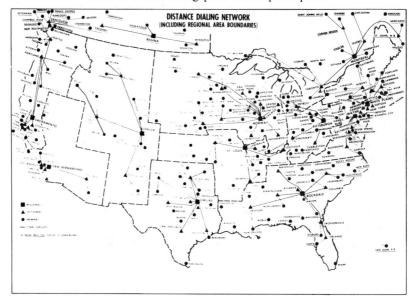

FIG. 4-2. Distance dialing network.

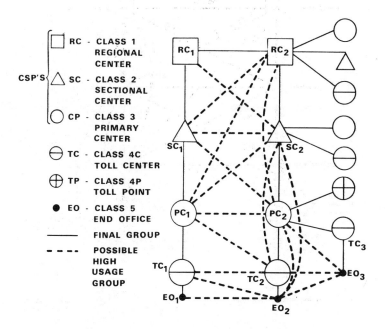

FIG. 4-3. Switching plan (routing pattern).

be established in the toll hierarchy, summarized in the *1-level limit rule:*

> The switching *function* performed for the first routed traffic by the switching system at either end of a high-usage trunk group *may be of the same class number of switching function, or may differ by at most one.*

The key word in the above rule is *function.* A high-usage trunk group may be established, for example, between a class 5 end office and a class 1 regional center for that traffic for which the regional center is performing a class 4 function (i.e., traffic destined for an end office homing on the regional center).

An example of the routing for a call that might have been placed some years ago from Bell Labs, Holmdel, New Jersey, to the State Department in Washington, D. C., is illustrated in Figure 4-4.

After reaching the class 3, Crossbar Tandem office in Asbury Park, the high-usage trunk groups to the class 3, 4A and class 2, 4A offices in Washington are tried in succession, followed by the final trunk group to the Newark 4A office. If the first groups are busy and the latter group has an idle

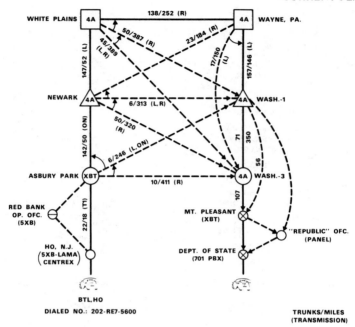

FIG. 4-4. Routing example.

trunk, the call advances to the Newark, class 2, 4A office, where the trial order is indicated by arrows. If the call reaches the descending hierarchical chain, a similar high-usage-to-final group trial ordering will apply.

The route consisting entirely of final groups is referred to as the *backbone* route.

It should be noted that the maximum number of trunks connected in the final route chain from class 4 office to class 4 office must not exceed seven for an interregional call in order to meet transmission impairment limitations. Adding the trunks to the class 5 offices at each end yields a maximum of nine trunks. Most connections involve one or two intertoll trunks with an average of 1.4, or 3.4 in all.

Recently, to allow implementation of new CAMA, TSPS and tandem arrangements, a new class of office has been introduced which may be interposed between class 4 and class 5 offices, and is designated a class 4X office. It may perform certain class 4 and class 5 functions, but is ideally transparent to the network. A class 4X office is termed an *Intermediate Point.*

4X offices must be essentially nonblocking in order to maintain the grade of service established before their introduction.

During the 1980's, a conversion to Dynamic Non-Hierarchical Routing (DNHR) will take place. This routing concept will effectively reduce the number of hierarchical levels to two, using the power of stored-program machines to dynamically change the routing strategy as a function of time and loading.

4.3 CHARGING

Charging is clearly an essential function of telephonic activity, since most of the operating company revenues are derived therefrom.

Periodic charges (usually monthly) are established by tariffs and implemented by accounting and billing procedures; they do not involve switching system facilities, and are not viewed as *charging features*. In very early times, all charging was on a flat-rate basis, the availability of the service being viewed as more significant than the number, duration, time of day, or distance of placed calls. Though the charging policies have changed radically for toll calls and for local calls for some business customers, the majority of residential customers still enjoy flat-rate billing for local calls (this situation will not continue, as will be discussed shortly).

Message-rate charging is on the bases of time and distance, and is usually one message unit for a call within a local zone, and proportionately greater numbers of message units for more distant (though still *local*) zones (see Figure 4-5); billing for such calls is on a bulk, nonitemized basis. Calls beyond the local area are considered *toll* calls and are individually itemized and charged for.

In some offices, charges are recorded in electromechanical *message registers* (see Figure 4-6) which are photographed in large groups once a month, and the information obtained is processed manually. In others, automatic *ticketers* print charges directly upon paper tickets. In offices with early AMA (Automatic Message Accounting) equipment, the charging information is punched on paper tape (newer versions employ magnetic tape) which is changed daily (at approximately 3 a.m.) and dispatched to an accounting office where it is automatically processed (see Figure 4-7). There is a requirement imposed by the regulatory commissions for retention of the original recording medium for a period of a month after the events are recorded.

The most modern AMA arrangements utilize data links to transport calling information to centralized revenue accounting offices.

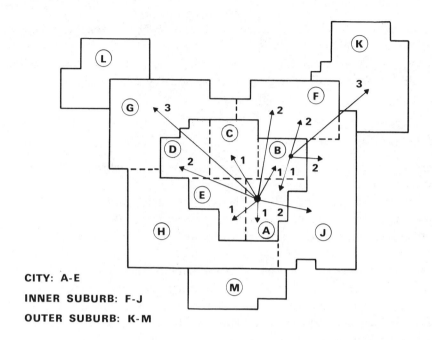

CITY: A-E
INNER SUBURB: F-J
OUTER SUBURB: K-M

FIG. 4-5. Zoning.

FIG. 4-6. Message registers.

FIG. 4-7. Paper tape recorder.

When an end office's size or makeup is such that (local) automatic accounting means are not justifiable, it may home upon an office with CAMA (Centralized Automatic Message Accounting) facilities. As is true for AMA, the directory number of the customer to be billed must be known, but this information is much more difficult to obtain for transmittal in the case of earlier electromechanical equipment, than to record locally. The identity of the originating line is readily available in implicit form in such an office because the switch train is set up to it and recording means (such as dedicated message registers) can be triggered easily via this route, but the explicit translation from equipment terminal to directory number for transmission to a distant CAMA office is possible only with the addition of ANI (Automatic Number Identification) equipment. This equipment generally applies a tone to the sleeve lead at the trunk side of the network and searches (while counting) for that tone at the line side. It was necessary only for Step-by-Step, Panel, and No. 1 Crossbar, since No. 5 Crossbar and later systems have the built-in ability to provide the calling party's directory number. Another version of CAMA known as *operator identified CAMA* employs operators to query the caller from an office without ANI for his directory number.

The operating companies have been caught in an economic squeeze in recent years between rising costs on the one hand and relatively static local telephone rates on the other. The rate commissions have indicated that their reluctance to grant rate increases would be substantially less where such increases were tied to usage level (there is increasing consumer sentiment that charges should be in proportion to usage rather than having light users in effect subsidize heavy users via flat-rate charging).

Rising costs for local service have not been due only to inflationary increases; users have been changing their calling habits, increasing their usage. Most lines are under a flat-rate charging program for local service, so increased usage does not result in increased revenues. Changing flat-rate billing to message-rate billing would correct this situation.

Other avenues of usage sensitive charging include charging for directory assistance and off-hook line (this would imply charging for false starts, permanent signal, and busy as well as dialing time and ringing time, etc.). There is no plan at present to pursue the latter.

Directory assistance has been demanding ever greater numbers of operators, and predictions indicated accelerating rates of growth until charging for this service began (see Figure 4-8).

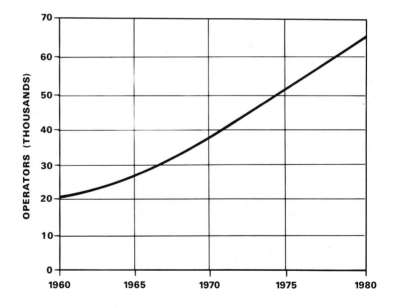

FIG. 4-8. Free directory-assistance operator requirements.

Before charging began, fifty percent of these calls were made by only 10 percent of the customers (see Figure 4-9), 80 percent of the calls were made by less than 20 percent of them, and 45 percent of the customers used the service very rarely.

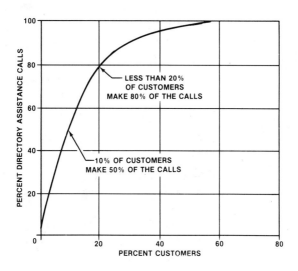

45% CUSTOMERS RARELY USE DIRECTORY ASSISTANCE

FIG. 4-9. Directory-assistance users.

An increasing proportion of these calls were for listings appearing in the local directories. Charging for this service has been introduced on the basis of a nominal allowed number of free assistance calls for numbers that are local or in the numbering plan area, with a charge for each additional call. Assistance calls from pay phones and from phones used by the handicapped remain free.

Cincinnati Bell received approval from the Public Utilities Commission of Ohio for a trial beginning March 3, 1974, allowing three free directory assistance calls per month, with a charge of $0.20 for each additional call. There is no charge for interstate toll directory assistance or for calls for the service from coin phones. Many other companies are now charging for directory assistance.

Provision of message-rate charging capability is hampered by the older equipment (Step-by-Step and, until recently, Panel), which is relatively inflexible in its charging capability; however, Panel equipment has been eliminated, and equipment upgrading the charging capability of Step-by-Step can be added.

4.4 SIGNALING

Signaling (see Figure 4-10) may be divided into two categories on the basis of the entities to be signaled between: station-office or office-office (interoffice).

	SUPERVISORY	*ADDRESS (PULSING)*
Station		
DC	Loop	Dial
AC	SF	Touch-Tone®
Interoffice		
DC	Reverse Battery	Dial
	High-Low	Revertive
	Wet-Dry	
AC	In-Band (SF)	Multi-Frequency
	Out-of-Band	Frequency-Shift
	Common Channel	

FIG. 4-10. Signaling.

Two natural subcategories in both cases are classified upon the nature of the information signaled: supervisory information or address (number) information. A further classification can be made on the basis of the nature of the signal (ac or dc, etc.).

The substantial number of types of signaling techniques reflects the evolution over the years in switching machines (which strongly influence the possible signaling means) and signaling technology, constrained by the need for compatibility. As a general rule, the dc signaling techniques are of earlier vintage than the ac techniques, and higher speed techniques also tend to be newer. Essentially every technique ever applied in the Bell System still survives to a greater-or-lesser degree.

Figure 4-11 details the signaling involved in a typical connection. Initially, the calling and called customers' lines, as well as some trunk destined to be selected, are idle and signal On Hook, the lines via drawing no current, the trunk by transmitting a Single Frequency (SF) signal. When the calling

customer originates, his line draws current, and when a signal reception means
is connected, it returns Dial Tone. The customer dials, either with a rotary dial
or a Touch-Tone pad, and after some processing, the control searches for and
seizes the trunk. The destination office will return a Delay Dial signal until it
has connected a signal reception means, then will interrupt the SF tone and
send a Start Dial signal. The originating office will send the Directory Number,
and the destination office will check the called customer's line for busy, then
will ring the line while returning Audible Tone to the caller. When the called
line answers, Off-Hook will be transmitted from the destination office, and the
path will be completed. After one or the other of the parties hangs up, the
paths will revert to their idle signaling states.

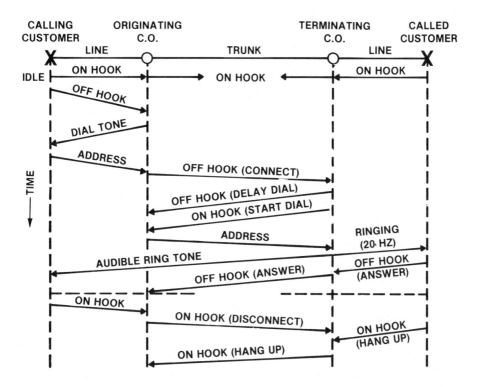

FIG. 4-11. Signaling for a typical connection.

Station-Office Signaling

Address signaling in the case of station-office communication takes place only from a station to a central office (not the reverse) and consists of rotary dial and TOUCH-TONE. The rotary dial breaks the dc loop a number of times equal to the dialed digit (with 0 interpreted as 10) at a 10-pulse per second rate. The TOUCH-TONE equipped station transmits one-each-out-of-two-groups-of-four frequencies (two frequencies; see Figure 4-12) for each dialed digit.

HIGH GROUP (HZ)

		1209	1336	1477	1633
	697	1	2	3	SPARE
LOW GROUP (HZ)	770	4	5	6	SPARE
	852	7	8	9	SPARE
	941	*	0	#	SPARE

FIG. 4-12. Touch-tone frequencies.

Supervision is normally on the basis of dc loop current.

There are many other forms of signals that are also exchanged between a station and its central office (see Figure 4-13).

Interoffice Signaling

Conventional Techniques

The means used for signaling between offices are a strong function of the office types involved (see Figure 4-14) and are rather more varied in their philosophies than station-office signaling. The older office types were incapable of signaling in other than their standard language, while the more modern machines are multilingual and able to signal to any other machine. These include:

Address

> *Dial Pulse* (DP) - A usually repeated version of customer dialing generated at a Step-by-Step office.

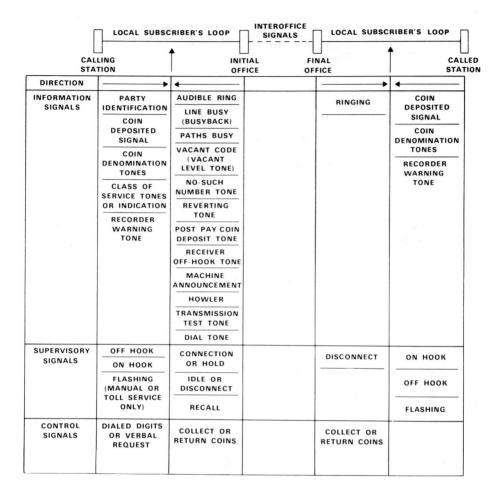

FIG. 4-13. Station-office signaling.

Key Pulse (KP) - Operator-keyed signaling technique (varieties range from DC to MF).

Revertive (RP) - A feedback type of signaling that was required by Panel offices.

Panel Call Indicator (PCI) - A means for providing number information to a manual operator via combinations of polarized pulses automatically interpreted and displayed.

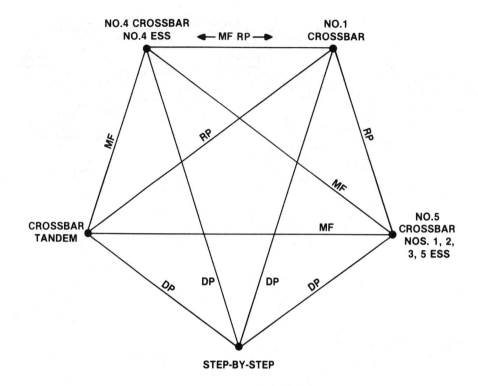

FIG. 4-14. Interoffice signaling.

Multifrequency (MF) - A 2-out-of-6 frequency technique; the most commonly used today (Figure 4-15).

Frequency Shift Pulsing (FSP) - A relatively new, high-speed technique that transmits signals shifted ±100 Hz about 1170 Hz.

Relative speeds are given in Figure 4-16. (Note that a digit is equivalent to an average of 5 pulses.)

Supervision (Figure 4-17)

Reverse Battery - A dc technique which signals forward (a seizure) by closing the loop, and backward (answer) by reversing the applied potential; preferred method for modern system trunks.

DIGIT	CODES	FREQUENCIES (Hz)
1	0,1	700, 900
2	0,2	700, 1100
3	1,2	900, 1100
4	0,4	700, 1300
5	1,4	900, 1300
6	2,4	1100, 1300
7	0,7	700, 1500
8	1,7	900, 1500
9	2,7	1100, 1500
0	4,7	1300, 1500
KP	2,10	1100, 1700
ST	7,10	1500, 1700

FIG. 4-15. Multifrequency signals.

DP : 10 or 20 Pulses/Sec.

RP : 27-30 Pulses/Sec.

PCI : 3 Digits/Sec.

KP : ≈ 4 Digits/Sec.

MF : 7 Digits/Sec.

FSP : 33-1/3 Digits/Sec.

FIG. 4-16. Relative signaling speeds.

High-Low - A dc technique utilizing marginal current changes for signaling.

Wet-Dry - A loop-like dc technique with battery applied at the called end (wet) until answer, upon which it is removed (dry).

Single Frequency (SF) - An inband, AC technique using a single frequency (2600 Hz) for signaling over carrier facilities.

Out of Band - An out-of-band, AC technique using a single out-of-band frequency (3700 Hz) for signaling over carrier facilities.

Inband signaling techniques suffer from potential voice synthesis of signals that can create false disconnect (talkoff) or mask a signal (talkdown), which can cause undesired situations such as charging for a free call.

Out-of-band signaling restricts the voiceband somewhat, but is free of talkoff and talkdown problems and can be used during customer conversation.

- ● **DC**
    Reverse Battery (Preferred Method)
    High—Low
    Wet—Dry

- ● **AC**
    Single Frequency
    Out-of-Band

FIG. 4-17. Interoffice supervision.

COMMON CHANNEL INTEROFFICE SIGNALING

Signaling between telephone offices has conventionally been performed over the speech-path trunks themselves. This procedure has the advantage of preventing assigning a bad trunk for a speech path, because the signaling itself would not be properly handled by a trunk having severe problems. However, there is the possibility of producing unintended signals when the signaling is in-band, and out-of-band signaling also has some disadvantages. For these and other reasons, a system known as Common Channel Interoffice Signaling (CCIS) is being deployed. This is a scheme for permitting communication between the controls of central offices via a private channel, indicating the speech-path connections to be made on trunks between the two offices (Figure 4-18).

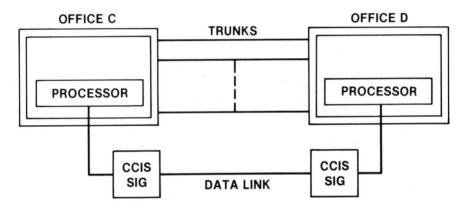

FIG. 4-18. Common Channel Interoffice Signaling.

In a sense this scheme is a "throwback" to one of the very early systems used for handling manual interoffice traffic, called the *order wire* system, where the operator in one office would call the operator in another office over a private line, indicating which trunk to use.

Initial deployment of CCIS is aimed primarily at communication between toll offices, though it is expected soon to be applied to offices lower in the hierarchy.

Its advantages are:

1. Increased signaling capacity and versatility.

2. Reduced post-dialing delay.

3. Prevention of fraudulent customer control (blue boxes).

4. Elimination of *talk-off*.

5. Elimination of per trunk equipment.

The initial system used a 2400-bit-per-second channel capable of handling some 3000 trunks. The channel is duplicated for reliability. Extension to a 4800-bit-per-second capability was introduced in 1982, and expansion to 5600-bits-per-second is expected soon.

At present, *nonassociated CCIS* (Figure 4-19), whereby a separate signaling network is established, with signals being routed through Signal Transfer Points (STP), is the primary form implemented.

The first CCIS facility was put in service in May, 1976, between Madison, Wisconsin and Chicago, Illinois, and the entire (toll) complement of 20 STPs (2 per region) has been deployed.

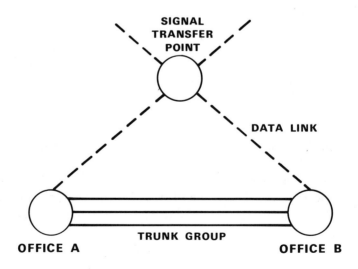

FIG. 4-19. Non-associated CCIS.

There has been international activity in CCIS as well. Standards have been set for such capability which, though compatible, are not identical to those being applied within the Bell System. A field trial was run (from 1968 to 1972) using CCIS in communication between the United States, Australia, and Japan, and the results were quite favorable.

4.5 NUMBERING PLAN

Domestic

The administration of a numbering plan for the nation would appear to be a simple task; 10 digits are routinely dialed for toll calls, and 10 decimal digits clearly yield 10^{10} unique numbers, some two orders of magnitude greater than the present number of lines in service in the United States.

However, various constraints with historical, political, and practicability bases complicate the problem. First of all, it is important to strive for a "uniform" numbering plan so that a mobile population will not have to apply different dialing procedures in different geographic areas. Next, it is desirable to minimize the number of digits that must be dialed for most calls; thus 7 digits are used for local dialing, while 10 (or 11) are used for toll calls. All-number calling was postponed as long as possible (and still is not universal). Soon it will be necessary to add yet another digit (a leading 1) to the string required for station-to-station toll dialing (this is already necessary in many areas).

The entire United States, Bermuda, the Caribbean Islands, Canada, and a portion of Northwest Mexico have been partitioned geographically into *Numbering Plan Areas,* (NPAs) each of which is assigned a distinct 3-digit *Area Code* (see Figure 4-20). Calls between NPAs, in general, require the dialing of the NPA code in advance of the called customer's 7-digit number. Calls originating and terminating within the same NPA require the dialing of only seven digits.

The distance dialing code syntax (see Figure 4-21) is as follows:

1. A 3-digit NPA code or area code with digit combination restrictions, followed by

2. A 3-digit office code with similar restrictions, followed by

3. A 4-digit station code with no combination restrictions.

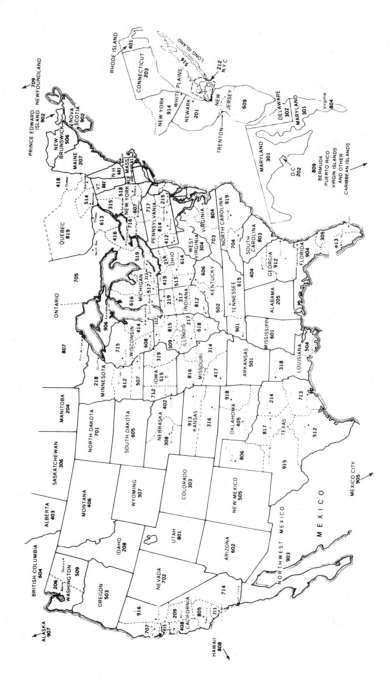

FIG. 4-20. Numbering plan areas.

	AREA CODE	LISTED DIRECTORY NUMBER	
		OFFICE CODE	STATION CODE
PREVIOUS	N0/1 X	NNX	XXXX
PRESENT	N0/1 X	NXX	XXXX
ULTIMATE	1NXX	NXX	XXXX

X: 0 TO 9
N: 2 TO 9

FIG. 4-21. Syntax of distance dialing.

The restrictions have existed to preserve as long as possible the mutually exclusive status which has been enjoyed by the NPA and office codes and to reserve a few codes for special purposes. The present restriction dictates that the leading digit of both NPA and office codes, and the second digit of offices codes, cannot be 0 or 1.

In early 1974, the codes possible under the present NPA restriction were exhausted (in the 213 NPA), and identical NPA and office codes became possible. It has therefore become necessary to distinguish between a 7- and a 10-digit call by other means. What had been planned (but was reconsidered) was to introduce a 4-second delay after the seventh digit. If no additional digit were dialed in that interval, the switching machine would conclude that it was a 7-digit call. It appears likely that a "1" (or "0") prefix to 10-digit calls will be adopted instead.

The first such situation arose in Los Angeles, where the usable office codes have been exhausted (chiefly because of the inefficient code-fill of the many small Step-by-Step offices), and a leading 1 has been required since July of 1973 for all toll calls, because early in 1974 one of the Canadian NPA codes (613) began use as an office code in L.A.

International DDD

In 1958, negotiations began with some European telephone administrations exploring the possibility of international dialing. As a result, in 1963, *Operator Distance Dialing* (ODD) was introduced on circuits from the United States to England and Germany. Currently, ODD is used on circuits to certain other European countries, Australia, Japan, and elsewhere (Figure 4-22).

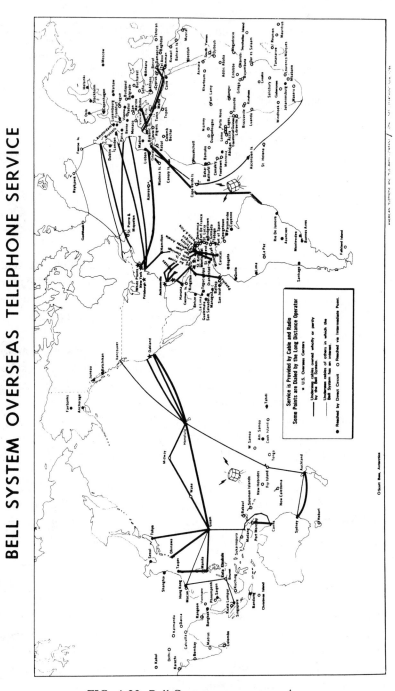

FIG. 4-22. Bell System overseas service.

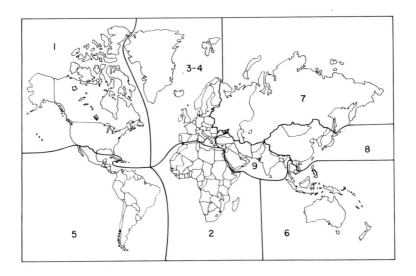

FIG. 4-23. World numbering zones.

The organization working to improve international communications is the International Telephone and Telegraph Consultative Committee (CCITT) of the International Telecommunications Union (ITU). The world-wide numbering plan developed by this organization provides each customer with a unique world telephone number that consists of a country code prefix to the national code and may not exceed 12 digits. The world is divided into zones (see Figure 4-23) with each country assigned a code which may have one, two, or three digits, e.g., the USSR number is 7, Belgium's is 32, and Portugal's 351. The world telephone number of the author at Bell Laboratories, Naperville, Illinois, for example, is 1-312-979-3296 (after dialing the access code).

The plan for providing International DDD is based upon ESS offices and electromechanical offices with TSPS capability. Local common control offices where this offering will be made must be modified to accept the international access codes plus up to 12 digits. The international access codes are 011 for station-to-station and 01 for calls requiring operator assistance.

As was true even before ODD, calls will be routed to one of seven overseas "gateway" offices, which act as the interface for overseas traffic.

4.6 EXPANDED 911

The 911 universal emergency number was first made available in 1968. Since then over two-hundred and fifty 911 systems have been placed in service, and others are in various stages of planning and implementation. Implementation requires the establishment of an Emergency Service Bureau (ESB) to which 911 calls are routed. Basic service is provided at two levels, the lower of which comprises:

Forced Disconnect   To eliminate nuisance use and intentional jamming.

Tone Application   To allow differentiation between parties who have abandoned before pickup, and those who are on the line but cannot speak.

At additional charge, available features are:

Called Party Hold   To allow the ESB to hold up the connection for manual trace regardless of the calling party's switchhook status.

Ringback   To allow the ESB to ring a telephone being held; this feature requires called party hold capability.

Switchhook Status   To allow monitoring calling party status.

Various agencies have collectively requested certain enhancements of the basic service:

Automatic Number Identification

Automatic Location Identification

Selective Routing (Routes the call automatically to the most geographically appropriate ESB).

The Office of Telecommunications of the U. S. Department of Commerce opened a 911 information center to aid local officials in obtaining current data on the service. This center's existence is expected to encourage wider adoption of 911 service and discourage the concept that separate numbers are desirable for separate agencies.

A significant difference in the charging policy is necessary for expanded 911 service. Because of the nontrivial volume of additional equipment required, special charges must be levied upon the agencies requesting additional features beyond the basic service.

4.7 ADVANCED MOBILE PHONE SERVICE (AMPS)

As the result of a relatively recent FCC ruling, a new portion of the communication spectrum has been made available for common carrier mobile services. This band is in the range from 806 megahertz to 881 megahertz. AT&T, as well as a number of other commercial communications equipment producers, vied for various portions of this spectrum.

The system proposed by Bell Telephone Laboratories is one which makes very efficient use of the frequencies allocated. Previous allocations of frequencies for mobile radio service are so meager that even in large metropolitan areas the number of available channels were of the order of ten, so that the number of customers who could be reasonably served in such an area was extremely small. The proposed system, called AMPS (Advanced Mobile Phone System), uses 64 megahertz of the band and is capable of serving numbers of customers several orders of magnitude larger than that previously possible.

The AMPS cellular scheme utilizes a system of antennas placed in a triangular pattern at corners of a (conceptual) hexagon. This technique makes use of the *capture* properties of frequency modulation, the approach being to allow multiple usage of the same frequencies in relatively close proximity (Figure 4-24). One aspect of this scheme is the necessity of determining the location of a vehicle within a hexagonal pattern so that it may be properly

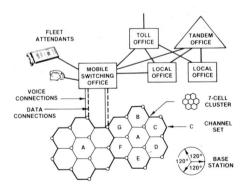

FIG. 4-24. Advanced Mobile Phone Service.

served by the triad of antennas within that pattern. Another is that the frequency used for transmission to and from the mobile vehicle is altered as it passes from the geographic area of one hexagon to another, commands being sent to the receiver in the mobile unit to bring about this frequency alteration. With this scheme, frequencies may be used many times over within a metropolitan area, yielding extremely good spectrum efficiency (see Figure 4-25). The switching function, however, becomes quite complex, necessitating a stored-program machine control to provide the rapid response to changing conditions necessary for effective use of this system.

In July, 1978 in Chicago, Illinois, testing of the mobile telephone cellular system began, and a service test offering to selected customers began a six-month trial in January, 1979.

The success of the trial fostered FCC approval of the approach, and AT&T formed a new subsidiary, Advanced Mobile Phone Service, Inc. to provide the service along with several other telecommunications companies. By the June, 1982 deadline it had set, the FCC had received 196 cellular mobile radio license applications.

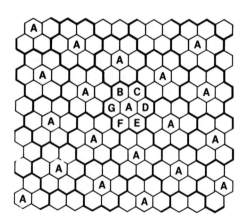

FIG. 4-25. AMPS channel reuse.

EXERCISES

1. If you were called upon to make the decision, above what size would you require a PBX to have duplicated equipment for reliability?

2. Usage-sensitive charging is under consideration throughout the telephone industry. What would be your opinion of charging a customer for all off-hook time, whether he is the calling or called party (including, e.g., call attempts to busy lines)?

3. It will have been noted that interoffice signaling is on a "link-by-link" basis, each office in the chain collecting information, then forwarding to the next office in the chain. What would be the effect of end-to-end signaling, where the originating office does all the signaling through intervening offices to the next one requiring information?

4. If, say, an asterisk were permitted to be used as the third digit of an NPA code (unlikely until 12-button TOUCH-TONE stations are virtually universal), how many additional codes would become available? (Assume otherwise "ultimate" syntax.)

5. What other alternatives to the 4-second delay and leading 1 might be considered as means for distinguishing 7- from 10-digit calls?

6. What other sources of revenue (other than telephone usage charges) has a telephone operating company?

READING LIST

Maintenance

R. W. Downing, J. S. Novak, and L. S. Tuomenoksa, "No. 1 ESS Maintenance Plan, *BSTJ* Vol. 43, pp. 1961-2019, September 1964.

A. L. Fleming, "Automatic Testing for Step-by-Step Offices," *Bell Labs Record,* Vol. 45, pp 153-156, 1967.

C. E. Germanton, "Alarm System for No. 5 Crossbar," *Bell Labs Record,* Vol. 27, pp 294-298, 1949.

The Hierarchy

Notes on the Network, AT&T, 1980.

Charging

G. V. King, "Centralized Automatic Message Accounting," *BSTJ*, Vol. 33, pp 131-1342, 1954.

F. K. Low, "Message Register Operation in No. 5 Crossbar," *Bell Labs Record,* Vol. 48, pp 404-408, 1950.

M. C. Goddard, "Coin Zone Dialing in No. 5 Crossbar," *Bell Labs Record,* Vol. 38, pp 106-108, 1960.

Signaling

C. Breen and C. A. Dahlbom, "Signaling Systems for Control of Telephone Switching," *BSTJ* Vol. 39, pp 1381-1441, November, 1960. Also reprinted as Bell System Monograph 3736.

Numbering Plan

Notes on The Network, AT&T, 1980.

911 Service

J. M. Shepard (AT&T), "911 - Its Present and Future," The National Association of Regulatory Utility Commissioners: Conference of Regulatory Engineers - Subcommittees on Communications, June, 1973.

Mobile Radio

Z. C. Fluhr and E. Nussbaum, "Switching Plan for a Cellular Mobile Telephone System," *IEEE Transactions on Communications,* Vol. Com-21, No. 1, November, 1973, pp 1281-1286.

James A. O'Brian, "Final Tests Begin for Mobile Telephone System," *Bell Laboratories Record,* Vol. 56, No. 7, July/August, 1978, pp 170-174.

CHAPTER 5

NETWORKS

5.1 GENERALITIES

The term *network* as used in Telephony has little relation to the term as used in electrical engineering. It refers instead to the means within a telephone office for interconnecting lines (for intraoffice calls), lines and trunks (for interoffice calls), and trunks (for tandem or toll traffic). In modern machines the power of a generalized switching network has been applied as well to the task of making connections to service circuits (e.g., TOUCH-TONE receivers, ringers, MF senders).

In SXS, the network and control are inextricably combined; in other systems (in general) they are separate and distinct (one exception is the rare end-marked network previously described).

The network may be regarded as the analogue of the manual switchboard and operator's arms, and the control that of the operator's mind (at least that part concerned with telephony). An operator can typically have access to 10,000 jacks (arranged in a 100-by-100 matrix), while modern small-motion switches have a much more modest *reach;* it is for the latter reason that multiple stage networks are necessary in modern systems.

Though most of the switching equipment in use today is of the space-division type implied above, there exist time-division networks and combinations of time and space division (e.g., No. 4 ESS), which will be discussed.

5.2 PRINCIPLES

In the earlier discussion of the need for centralized switching, it was shown that centralization not only reduced the number of transmission paths required, but also halved the number of crosspoints needed. The resulting number of crosspoints, however, would still be impractically large for any nontrivial office.

It is possible, in fact, to reduce the number of crosspoints using, e.g., Clos[1] networks (a nonblocking, full-access network structure studied by C. Clos, which will be considered in the next section), but the crosspoint count still leaves much to be desired.

The approach taken, therefore, in all practical, nontrivial space-division networks, is to engineer for nonzero blocking. Fortunately, crosspoint count is a sensitive function of blocking probability for a given traffic load, and is very materially less than for a Clos network for blocking probabilities of the order of .01 (compared to 0) for moderate traffic.

Consider the case of eight subscribers to be accommodated by a centralized switching mechanism (see Figure 5-1; the definition of the term *strict* as used in the figure will be given later).

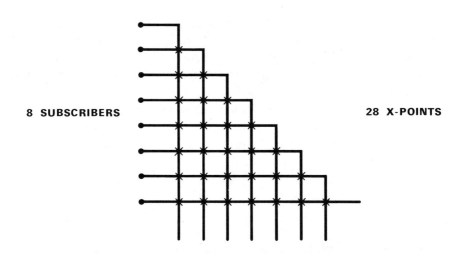

8 SUBSCRIBERS   **28 X-POINTS**

FIG. 5-1. Nonblocking full access (strict).

---

1. C. Clos, "A Study of Nonblocking Networks," *BSTJ*, Vol. 32 pp. 406-424, March, 1953.

The maximum number of conversations possible is four (half of the subscribers speaking to the other half). The crosspoint count is $(8 \times 7)/2 = 28$. In practice, it would be unlikely for all subscribers to be using the phone at once (the probability approaches zero as the number of subscribers increases), and it would be needlessly extravagant to provide the equipment necessary for such capability. Indeed, it is sufficient in practice to provide enough equipment (plus ϵ) to satisfy the average peak of traffic (usually of the order of 10 percent of the lines active).

Suppose that we decide to provide a maximum of two conversations among our eight subscribers; then we might use the arrangement of Figure 5-2, where we have reduced the number of crosspoints to 20, using three stages of switching.

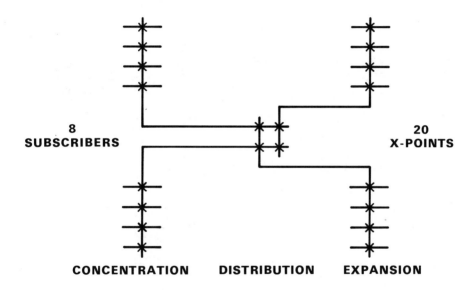

FIG. 5-2. Blocking full access.

The leftmost stage is a *concentration* stage, concentrating eight lines to two *links*. The center stage is the *distribution* stage, providing *full access* from any input to any output. The rightmost stage is an *expansion* stage, expanding two links to eight lines. (It happens that, for so few lines, a single-stage switch with fewer (16) crosspoints can do the same job; for practical numbers of lines, however, multistage networks are necessary for crosspoint economy.)

In practical, large networks, the same elements of concentration, distribution, and expansion are employed, but in a variety of configurations, usually employing a multiplicity of switching stages for each element and yielding more significant reductions in the number of crosspoints required relative to a nonblocking case.

It should also be pointed out that the number of wires that must be switched may be one, two, three, four, or six. One line was switched in the very early days of telephony (though soon discarded due to ground-return noise), but is being used (in a sense) for some modern electronic networks (though 2-wire transmission would still be used outside the office). Two wires (tip and ring), analogous to corresponding portions of an operator's plug (see Figure 5-3) are switched in local ESS offices.

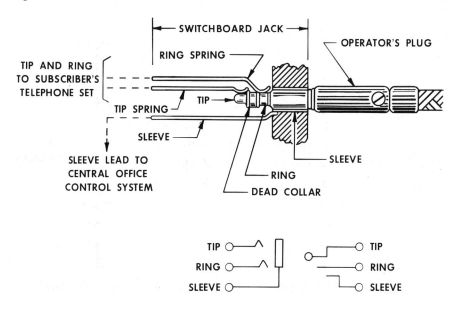

FIG. 5-3. Origin of network terms.

Three wires are used in the electromechanical offices, the third wire being called the *sleeve* in analogy to that part of the plug used by the operator to check for busy/idle status; it is used by the equipment for the same purpose and is replaced by a *network map* in the memory of ESS machines. Four wires (two, 2-wire paths) are switched in ESS offices handling toll traffic, to reduce the degradation otherwise experienced in long haul, multiply switched paths (due to the necessity otherwise for repeated use of *hybrids*[2]). Six wires (two 2-wire speech paths plus two sleeve leads) are switched in electromechanical toll offices.

Referring again to Figure 5-2, it will be seen that, though two simultaneous conversations can be carried on, and any single conversation is possible, not all pairs of conversations are possible. This is due to internal blocking or mismatching of available equipment. This phenomenon occurs as well in large networks.

This illustration is useful for pointing out another phenomenon. The likelihood of blocking in a network with so few inlets is quite high because of the "granularity" of the offered traffic. In a scaled-up version, though the concentration ratio remained the same, the larger numbers would "smooth" the traffic and substantially reduce the probability of blocking. Nonetheless, it is possible to experience undue blocking in the first stages of any nontrivial network if particularly heavy-use customers or groups thereof by chance find themselves sharing the services of the first switch. When such a circumstance occurs, it is necessary to perform a function known as *load balancing,* which entails moving the high-use lines to other, less heavily loaded switches.

5.3 NONBLOCKING NETWORKS

Some insight into networks with blocking can be gained by exposure to those without it. A nonblocking network is capable of guaranteeing a path from any idle inlet to any idle outlet irrespective of traffic.

Though up to this point we have assumed that the switching was amongst the universe of subscribers (i.e., each inlet was also an outlet or, alternatively, only intraoffice traffic was possible), from this point we will consider the more general case of switching a set of inlets to a set of outlets that may have no relation to the inlets.

A single-stage nonblocking network switching N inlets to N outlets requires N^2 crosspoints (see Figure 5-4), and this value serves as an upper bound upon the number required to attain this property.

---

2. Means for converting from 2-wire to 4-wire transmission and conversely.

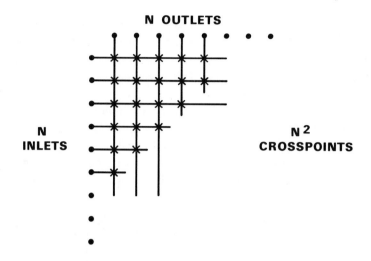

FIG. 5-4. Single-stage nonblocking network.

Clos considered multistage nonblocking networks, his general approach being to prevent *mismatch blocking* (blocking because, though the first and last stages are no problem, the intermediary stage [or stages] does not have an idle path that matches-up) by employing expansion in the first stage to assure sufficient links such that an idle matching path is always available. More specifically, consider Figure 5-5, a 3-stage network with N inlet ports and N outlet ports. Nonblocking arrays are used, rectangular in the first and last stages, and square in the center stage.

Clos' reasoning is quite simple (Figure 5-6); if a given inlet must be able to see a path through the center stage to a given outlet, then at most it must compete with the other $n - 1$ inlets in its array and the $n - 1$ other outlets in the (third-stage) array serving its outlet of interest. Provision of at least one more second-stage path than could be thus occupied assures nonblocking, for a net of $2n - 1$ paths. The number, k, of second-stage arrays (or switches) is therefore made equal to that number. The number of crosspoints can be calculated as follows:

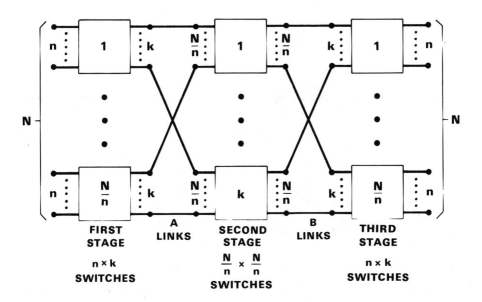

FIG. 5-5. Three-stage space-division network.

● **Cause of Mismatch Blocking:**

The $n-1$ Other Inlets to Source Switch and/or
the $n-1$ Other Inlets to Destination Switch
Preclude the Use of an Intermediary Switch.

● **Solution:**

Provide Enough Intermediary Switches for the
Worst Case Plus One, i.e.,

$$2(n-1)+1=2n-1$$

FIG. 5-6. Clos reasoning.

First stage: $\dfrac{N}{n}(n \times k) = kN$.

Second stage: $k\left(\dfrac{N}{n} \times \dfrac{N}{n}\right) = k\,\dfrac{N^2}{n^2}$.

Third stage: $\dfrac{N}{n}(n \times k) = \dfrac{kN}{k\left(2N + \dfrac{N^2}{n^2}\right)}$.

For a total of $k\left(2N + \dfrac{N^2}{n^2}\right)$.

For a Clos Network, $k = 2n - 1$, and

$$\textit{Total Crosspoint Count} = (2n-1)\left(2N + \dfrac{N^2}{n^2}\right) .$$

The optimum value of n can be found by differentiating with respect to n and setting equal to zero, yielding a cubic in n whose solution may be approximated for large N by:

$$n = \sqrt{\dfrac{N}{2}} .$$

Figure 5-7 tabulates comparative crosspoint counts for various Ns; note that the three-stage Clos network has a worse count for small N than a single-stage square array.

The Clos approach can be applied as well to produce multistage networks with more than three stages by expanding the center stage into a three-stage Clos network, which causes the network to become five-stage. This process can be iteratively applied until the center stage becomes trivially small. In fact, the number of stages in a Clos network with the fewest crosspoints increases with the number of inlets (and outlets). This phenomenon carries over as well into

	NEAREST INTEGER		CROSSPOINT COUNT	
N	**n**	**k**	SINGLE STAGE (N^2)	3 STAGE CLOS
16	2	3	256	288
27	3	5	729	675
60	5	9	3,600	2,376
105	7	13	11,025	5,655
200	10	19	40,000	15,200
8,192	64	127	67,108,864	4,161,539

FIG. 5-7. Nonblocking network crosspoint count.

blocking networks.

As an example of a Clos network (Figure 5-8) consider again the 8-inlet by 8-outlet case ($N = 8$) and choose $n = 2$. Then:

$$k = 2n - 1 = 3,$$

resulting in four 2-by-3 switches in the first and third stages, and three 4-by-4 switches in the second stage. The crosspoint consumption is 96. (The Exercises include an $n = 4$ Clos network.)

For more than 11 stages, a complex structure due to Canter is more economical than Clos networks, and for still larger networks, further improvement is possible with a structure due to Bassalygo and Pinsker. Practical networks rarely grow to so many stages, however.

5.4 BLOCKING NETWORKS

Blocking networks provide more economical implementations than nonblocking structures, but they must be carefully designed for satisfactory performance.

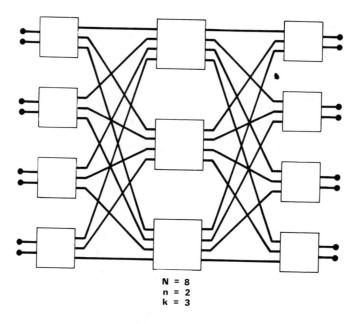

FIG. 5-8. Three-stage Clos network.

Two well-known analytical techniques exist, the Jacobaeus and Lee methods, only the second of which will be described.

The technique due to Lee permits calculation of blocking probabilities to a good approximation. It entails the use of a network graph that enumerates all possible paths leading from any inlet to any outlet (see Figure 5-9).

For a three-stage network with k intermediary, second-stage switches, there are k possible paths between a given inlet/outlet pair (see Figure 5-10). If the occupancy (or probability of being busy) of an inlet is p, the probability of a link being busy under the assumption that the traffic is uniformly distributed over the links (a situation only approximated in real networks) is

$$Prob.\ Link\ Busy\ =\ \frac{pn}{k}\ ;$$

n/k can also be called the *concentration ratio*. The probability of a path consisting of two links being busy is then (under the additional assumption that the link busy-idle states are independent)

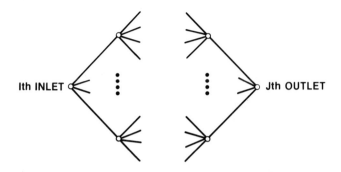

FIG. 5-9. Lee's method.

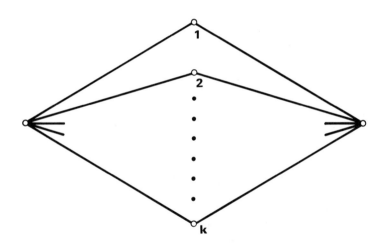

FOR INLET OCCUPANCY p, PROB. LINK BUSY $= \dfrac{pn}{k}$

PROB. PATH BUSY $= 1 - \left(1 - \dfrac{pn}{k}\right)^2$

PROB. ALL PATHS BUSY = PROB. BLOCKING $= \left[1 - \left(1 - \dfrac{pn}{k}\right)^2\right]^k$

FIG. 5-10. Three-stage network example.

$$\text{Prob. Path Busy} = 1 - \left[1 - \frac{pn}{k}\right]^2,$$

and the probability of all paths being busy is

$$\text{Prob. All Paths Busy} = \left[1 - \left[1 - \frac{pn}{k}\right]^2\right]^k,$$

which is also the probability of blocking.

Consider the example of Figure 5-11, which depicts the network graph for our earlier example (Figure 5-2).

GRAPH:

PROB. BLOCKING $= [1 - (1 - .4)^2]^1$
$= .64$

FIG. 5-11. Blocking: eight-customer example.

Since there is only one path from an inlet to any outlet, a particularly simple graph results. The probability of blocking is therefore given (assuming $p = 0.1$ and noting that the concentration ratio is 4) by

$$\text{Prob. Blocking} = \left[1 - (1 - 0.4)^2\right]^1$$

$$= 0.64,$$

which is a, perhaps, not unexpectedly high value in view of the primitive (though economical) network. As a further example, consider Figure 5-12, which depicts an 8-inlet by 8-outlet network with concentration ratio of 1 and $n = 2$.

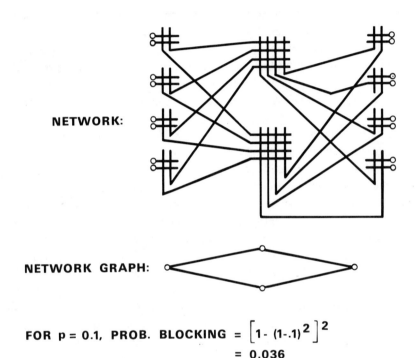

NETWORK:

NETWORK GRAPH:

FOR p = 0.1, PROB. BLOCKING $= \left[1 - (1-.1)^2 \right]^2$

$= 0.036$

FIG. 5-12. 8×8 network example.

The graph now has two paths and the blocking (again for $p = 0.1$) is

$$Prob.\ Blocking = \left[1 - (1 - 0.1)^2 \right]^2$$

$$= 0.036,$$

which is a very much better value than the previous example, but has cost more than three times as many crosspoints.

A discussion of a more complex network structure will be postponed until Section 5.12.

5.5 SIMULATION TECHNIQUES

Nontrivial networks present extraordinarily complex computational problems to the analyst. The techniques discussed thus far are inadequate to the task of reasonable determination of network behavior. Further, they cannot take cognizance of any path-hunt technique other than random selection. In practice, therefore, new network structures are usually subjected to simulation.

Two general approaches can be taken to network simulation: static and dynamic.

The static approach (epitomized by NEASIM, the simulator first used in analysis of the network for No. 1 ESS) employs the principles of the Lee graph method. Given the average occupancies of the network links, it creates an instantaneous network environment with links (randomly) marked busy in the proper proportion to satisfy the average occupancy probabilities. It then attempts to find complete paths for a small number of call attempts, notes the number unsuccessful, then creates another environment, etc. When a sufficient number of *samples* are taken to satisfy the confidence interval desired, the probability of blocking may be calculated from the results.

The static approach, while it can be economical with respect to computer time, suffers some drawbacks. It cannot assess the effects of various path-hunt algorithms, it may introduce difficulties with concentrating switches, it may create unrealistic network conditions (e.g., a link may be marked busy where there are no corresponding busy links to complete a path), and its results become less reliable for smaller networks.

The dynamic simulation approach models the network's activity in as realistic a manner as possible. It employs a traffic-generating mechanism that offers traffic to a simulation model of the system at a level corresponding to that under consideration. It employs whatever path-hunt algorithm the real system contemplates using, and (in effect) sets up and tears down the paths as calls enter and leave the system, keeping track of failures to find a path (blocking).

Because most networks are designed for relatively low blocking (.01-.001), very many simulated calls must be handled before an occurrence of blocking may be observed; very long, expensive simulation runs are therefore often necessary. Further, after simulation begins, a substantial amount of traffic must be simulated before an equilibrium condition corresponding to the desired busy-hour traffic situation obtains; data taken during the start-up, *transient* period is not useful and must be discarded, and that portion of the simulation time is, in a sense, wasted. (It is possible to start the simulation with the network entertaining a status close to equilibrium conditions [perhaps from a previous simulation] to reduce the simulation expense.)

Analysis of the results of dynamic simulation is less straightforward than that for static simulation because of the correlation between the data (e. g., the probability of a call entering the system is higher if the previous call was blocked). The powerful tools of Analysis of Variance depend upon random sampling and are virtually paralyzed in the face of correlated samples. Nevertheless, analysis is possible using correlation reduction techniques and/or very long runs.

5.6 NETWORK TECHNOLOGY

Electromechanical

Open Contact

Open contact technology includes SXS, Panel, and X-bar switches, all of which are electromechanical. Contacts are exposed to the atmosphere in all of these systems (though dust covers are provided for SXS switches), and airborne dust and substances that tend to corrode the contacts are major problems, as is wear (chiefly for the gross-motion machines; X-bar switches have only minimal wear problems).

Improvements in contact materials over the years (e.g., semiprecious metal use) greatly eased contact problems, but burnishing and lubricating contacts is still necessary for the gross-motion switches, and constitutes a substantial maintenance load.

Each of these networks consumes power while holding up a connection (though magnetically and physically latching X-bar switches have been developed for a few systems).

Closed Contact

Ferreed:

The ferreed technology employed in No. 1 and No. 2 ESS is electromechanical, closed contact. The contact elements are reed switches that are caused to close or remain open depending on the sense of the remanent state of the square-loop magnetic material associated with them, which is capable of being switched at "electronic" speed.

The contacts are hermetically sealed within a glass capsule, eliminating the atmospheric contamination problem. They are somewhat less robust than some other contact technologies, however, in that they must be switched "dry," that is, they are not permitted to open or close while carrying current.[3] The further precaution is taken via control programs, to rotate usage, equalizing wear. No power is consumed in holding up a network path.

Remreed:

The remanent reed (remreed) switch element employs reed blades that are magnetically hard, replacing the external magnetic material, permitting a smaller, lighter, more sensitive, and less expensive switch and network (see Figure 5-13). This network type has supplanted the production of ferreed networks for No. 1 and No. 2 ESS, and is used for the No. 3 ESS network.

Electronic

There has been interest for many years in applying electronic componentry in the networks of switching machines. In very large offices, the majority of the office cost is concentrated in the network, and in all but the smallest offices its contribution to the office cost is substantial.

The major impediments to electronic networks have been:

1. Inability to carry ringing current,

2. Need for costly interfaces to the electrically hostile external environment, and

---

3. Crossbar contacts are also normally switched dry.

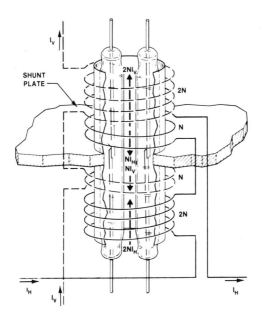

FIG. 5-13. Remreed crosspoint.

3. Need (in most cases) for controlled gain.

Space Division

The emphasis in electronic space division activity has historically been upon 4-layer diode technology (Figure 5-14), which provides storage as well as switched transmission capability; i.e., the crosspoint is electrically latching in that, once selected, it will remain conducting as long as the path's integrity is preserved. A number of inherent problems associated with the use of these elements have been solved (e.g., a rate-of-change dependent threshold effect and a path-sensitive attenuation problem), and these components have been applied in the 10A-RSS system (a remote switching vehicle).

Rugged electronic crosspoints have also been under study in various parts of the world (particularly in Japan and the U. S.) and have been applied in the concentrator of the No. 5 ESS.

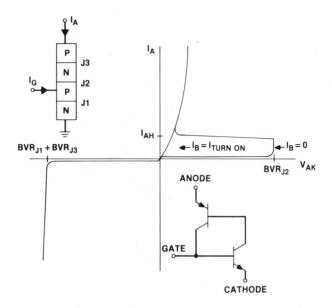

FIG. 5-14. PNPN characteristics.

Time Division

Early time-division networks (e.g., that of ESS 101) employed switches capable of preserving analog amplitude. The technology of modern Time-Division Multiplexing (TDM) methods tends to be gate-oriented and tracks gate technology; e.g., the Pulse-Code Modulation (PCM) networks for No. 4 and No. 5 ESS.

5.7 TECHNIQUES

Space Division

Space division is the name applied to the type of network discussed earlier in the chapter to distinguish it from the various time division networks. It emphasizes the fact that space or time (or both) may be used to separate information paths (as well as frequency which, though it has not yet found application in conventional switching systems, has been proposed for distributed systems).

Time Division

Time-Division Multiplexing makes use of the high speed readily available from modern electronic componentry to impress upon a single *highway,* a multiplicity of information paths, represented by their samples and separated by their time of appearance (Figure 5-15).

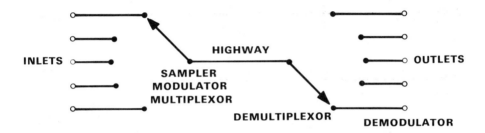

FIG. 5-15. Time-division multiplexing.

Figure 5-16 indicates the basic nature of the script commonly used in TDM techniques. In Pulse Amplitude Modulation (PAM), the amplitude of a periodic pulse is varied in accordance with that of the modulating signal (this technique has many of the properties of AM). Pulse Width Modulation (PWM), akin to Pulse Position Modulation (PPM), varies the width of the pulse (yielding many of the properties of FM). Pulse Code Modulation (PCM) provides a binary-coded numerical approximation to the sample amplitude. Delta Modulation provides, e.g., pulses when the modulating signal is increasing and none when it is decreasing (the frequency of pulses for this technique is usually much higher than that for the other methods). This method suffers from an inability to follow rapidly changing signals (slope overload), and adaptive variations upon the scheme are often used.

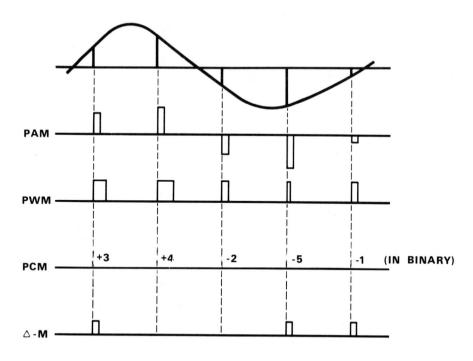

FIG. 5-16. Modulation scripts.

(The growing application of TDM in transmission will not be dealt with here, except for its influence upon switching.)

No. 101 ESS represented the first commercial Bell System offering of a TDM switching machine. The technique employed was Pulse Amplitude Modulation, with sampling at the rate of 12,500 per second. The Nyquist criterion[4] therefore limits the bandwidth to 6,250 Hz, well above 4,000 Hz, the "standard" telephone channel capacity (actually about 200 to 3,500 Hz).

No amplification was employed, rather a technique known as *Resonant Transfer* was used to reduce losses to a minimum; it was necessary therefore to "squirt" rather large current pulses through the common highway because the brief (2 microsecond) samples not only represented the amplitude of the signal, but also had to convey the energy of the sampled signal.

---

4. The Nyquist criterion dictates that the sampling rate must be at least twice the highest frequency component.

Among the difficulties which must be overcome with PAM are the effects of parasitic capacitance on the highway, which tends to retain a voltage "memory" of the last sample that, when added to the next successive sample, constitutes coherent noise or crosstalk.

Pulse Width Modulation (PWM) and Pulse Position Modulation (PPM), are similar, but differ from PAM in certain of their properties:

> They can be passed through logic gates without harm (a PAM signal would lose all its information).

> They can be amplified with high-efficiency class C amplifiers or logic gates.

Pulse Code Modulation (PCM) has been widely applied in transmission (where it originated), and its use is growing. PCM is valuable for such application because of its ability to "ruggedize" the signal against the corruptive influences that plague transmission using more traditional techniques. (It encodes the sample amplitude in a digital, multibit form.)

The use of T1 carrier is increasing at such a rate that it is advantageous to switch the digital signal directly instead of converting it to analog form for space-division switching, then reencoding it for retransmission. The switching networks for No. 4 and No. 5 ESS have this capability, as was described in Chapter 4.

Delta modulation is a variant of PCM whereby only the difference (rather than the absolute value) between a given sample and its predecessor is conveyed, typically via a single bit, effectively indicating the sign of the difference. This technique requires higher bandwidth, and may also be viewed as a limiting case of a more general technique known as differential PCM.

5.8 TRUNKING

Traffic Definitions and Characteristics

The traffic offered to a switching machine is a function of two parameters, the average rate of arrival of new call attempts (originations) and the average holding time of a call. The averaging period for origination rate is the *busy hour,* a one-hour period chosen to typify for a given office the annually recurring hour during which the offered traffic load is a maximum. Peak Busy Hour Calls (BHC) is the unit used for expressing the processing capacity of a switching machine's control.

The offered traffic load is expressed in *hundred call seconds* (CCS), and is the product of the number of calls and the average holding time, or the sum of the holding times of all calls under consideration. By informal convention, the units of CCS are often used to mean CCS *per hour*.

The average holding time multiplied by the number of calls placed per unit time is a measure of the traffic *intensity* and is expressed in *erlangs*. The erlang, named after an illustrious early contributor to traffic theory, is the traffic intensity equivalent to one call held for an entire hour, and is therefore equal to 36 CCS per hour (it is also equivalent, for example, to 36 calls held for 100 seconds each in a one-hour period, or a circuit occupied 100 percent for a full hour by any number of calls).

Put another way (see Figure 5-17), the traffic intensity in erlangs is the number of channels that would be sufficient to serve a given offered load in a one-hour period, if the load timing could be rearranged so that all the channels would be continuously busy. It thus constitutes a lower bound upon the number of channels to carry that traffic intensity.

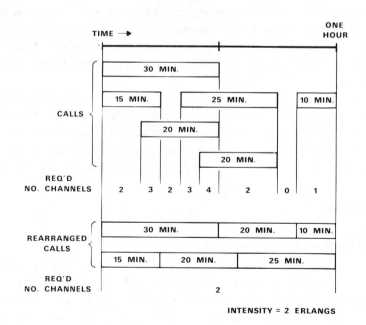

FIG. 5-17. Meaning of intensity in erlangs.

For example, an office serving 10,000 lines with an average origination rate of three calls per hour and an average holding time of 3 minutes (both typical) in the busy hour, would experience an offered traffic load of 30,000 BHC and 54,000 CCS per hour (or 1,500 erlangs). (From this point on, observing the convention, CCS will mean CCS per hour.)

Exhaustion of an office will usually occur because one or the other of the network or control becomes limiting; some offices, for example, tend to be limited by the number of BHC that can be handled by the control, while others tend to be limited by the network's capacity.

Observed offered traffic patterns are remarkably variable, being subject to time of day, season, location, and random events (see Figures 5-18 through 5-27).

Figure 5-18 shows the calling rates for 40 business subscribers in a New York City central office, arranged in order. Figure 5-19 shows the variations in holding time for a number of local calls (the behavior is virtually unchanged in cities as diverse as Newark, London, and Copenhagen). Figure 5-20 depicts the variation in traffic offered by the same 40 customers.

Figure 5-21 shows the hourly variation in residential and business calls. As might be expected, business calls peak rapidly at the start of the business day, and the effects of a fairly universal lunch hour can be seen. The residential behavior is slightly different, with an ill-defined lunch hour, but a peak in the early evening that may be explained as the traffic created by the business worker returned home.

Figure 5-22 shows an example of a central office in Brooklyn, New York, where the mixture of business and residential customers makes it difficult to define a busy hour.

Figure 5-23 records the behavior of traffic carried by an interoffice trunk group on days significantly separated in time.

Figure 5-24 shows the behavior of average busy hour calls on the same trunk group over several weeks.

Figure 5-25 depicts the seasonal variation in daily local calls. A dip can be seen to occur during the summer months when many vacations are taken.

Figure 5-26 shows the seasonal variation in daily toll calls. Note the rise during the summer months when the vacationers make calls.

Figure 5-27 records a typical pattern of traffic for a trunk group. It will be seen that while the instantaneous number of trunks in use varies randomly, there is a recognizable mean value, and a situation approximating statistical equilibrium exists.

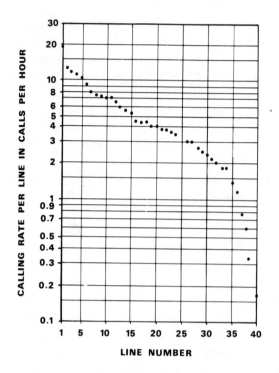

FIG. 5-18. Example of calling-rate variation.

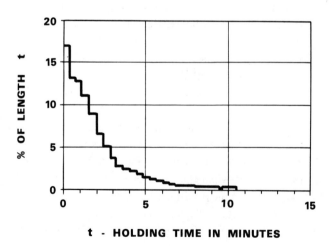

FIG. 5-19. Holding-time variation (local calls).

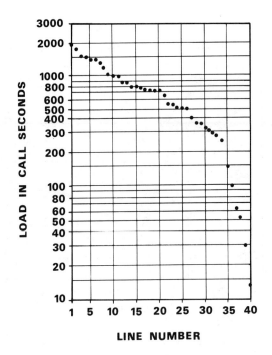

FIG. 5-20. Example of traffic-intensity variation.

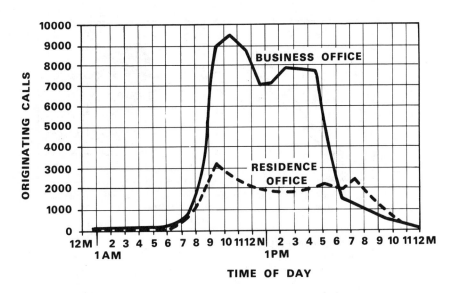

FIG. 5-21. Example of call variation.

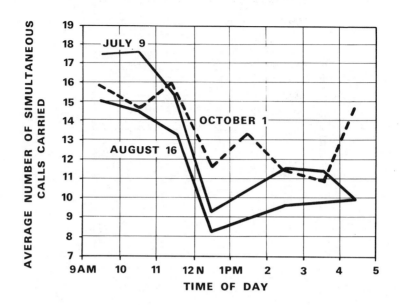

FIG. 5-22. CO Load-time relationship.

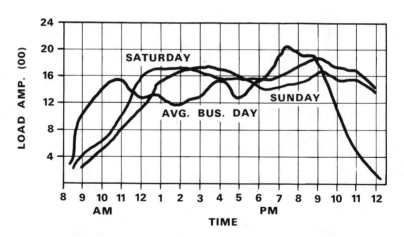

FIG. 5-23. Seasonal load variation.

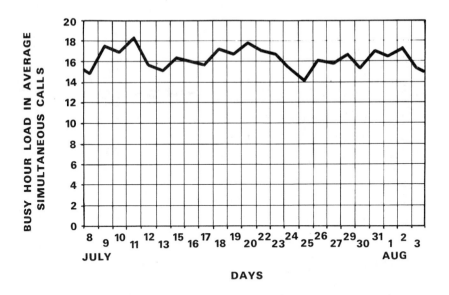

FIG. 5-24. Busy-hour load variation.

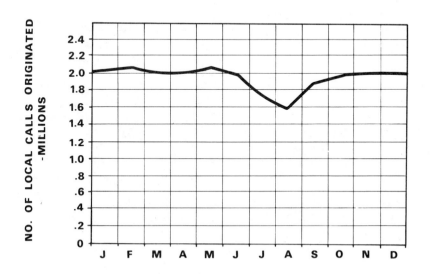

FIG. 5-25. Seasonal variation (local calls).

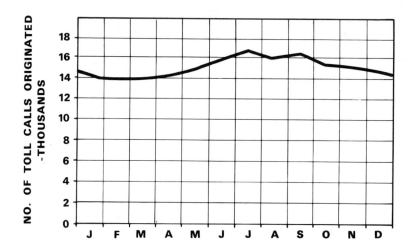

FIG. 5-26. Seasonal variation (toll calls)

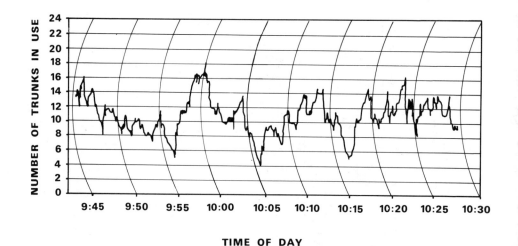

TIME OF DAY

FIG. 5-27. Busy-hour load distribution.

While traffic load is not totally unpredictable (Mother's Day and Christmas Day are dependably the peak yearly toll traffic days), occurrences such as a radio station offering a premium to the first 100 callers can cause a telephone traffic snarl to occur without warning.

It would clearly be uneconomical to engineer an office for an extremely rare peak of traffic (and indeed may be illegal),[5] but it is difficult to increase the traffic capacity of an office when the need arises (which tends to be much more frequent in metropolitan than in suburban or rural areas). It is necessary to predict the need for growth several years in advance (typically 2 to 5 years) because of the lead time in engineering, manufacture, and installation. Predictions are based on history, measurements of offered traffic in the office, judgments on the likelihood of population shifts (demographics) and changes in calling characteristics in the served area. Errors in such predictions can be near catastrophic in their effect upon service (e.g., the problems in Manhattan several years ago) or can result in *stranded capacity*.

Different switching machines respond in different ways to the traffic overload that will occur on occasion. To prevent an office from collapsing under overload, a *line load control* regimen is usually invoked, which typically treats lines served as comprising two classes, high priority lines (e.g., police, fire department, hospital, which are never denied service), and other lines. The latter are denied service for a time on a cyclic basis to reduce the offered load.

The determination of CCS per line is slightly complicated by the fact that intraoffice calls busy two lines, while outgoing and incoming calls busy only one. Similarly, BHC per line calculations must take incoming calls into account. The situation can be depicted in a traffic *H diagram* (see Figure 5-28). The H-diagram is based on the observation that the Divergence of traffic is 0. Intraoffice traffic is defined as a percentage of *originating* traffic. These diagrams usually are drawn with the assumption that originating traffic and *terminating* traffic are equal, and that incoming and outgoing traffic are equal, but any consistent proportions, including real measurements, can be used.

The CCS per line may be calculated as the sum of the originating and terminating CCS (which means that the intraoffice CCS is counted twice)

---

5. The regulatory commissions will not permit an office to contain substantially more equipment than is required for the immediately foreseeable peak busy hour; such a situation would be viewed as an improper investment of capital.

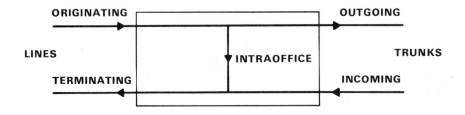

FOR ORIG. + TERM. CCS, INTRAOFFICE CCS COUNTED TWICE

FOR ORIG. + INC. CALLS, INTRAOFFICE CALLS COUNTED ONCE

FIG. 5-28. Traffic H-diagram.

divided by the number of lines. BHC per line is the sum of originating and incoming BHC (counting intraoffice calls only once) divided by the number of lines. Another factor taken into account is the effect of *false starts,* which may represent as much as 30 percent of the originations, and require less of the machine's processing capacity than correct attempts.

The mix of traffic offered to an office is a strong function of its location; a rural office will tend to have predominately intraoffice traffic because of the localized *community of interest,* while a metropolitan office will have relatively little intraoffice traffic. Suburban offices experience traffic behavior somewhere between these extremes.

Traffic Principles

The mathematics associated with traffic behavior are considerably simplified by certain assumptions commonly made. To properly apply the results, it is necessary to judge which of the assumptions most closely match the real-world situation at hand. We will begin with a set of general assumptions, draw conclusions upon them, then proceed with more specific assumptions.

We will assume that the sources (customers or terminals offering traffic to a trunk group or similar facility) are:

identical,

independent of other sources and the state of the system, and

collectively offering *random* traffic to the system.

Interpreting these assumptions, we conclude that, for sufficiently small time intervals, the probability of the arrival of a request for service is proportional to the time interval; i.e., if a time period t is partitioned into k equal intervals, where we choose k such that t/k is "sufficiently small," we may write

$$P(\text{Arrival in interval } t/k) = \lambda\, t/k,$$

where λ is the proportionality constant. Then the probability of no arrivals in that interval is given by

$$P(\text{No arrivals in interval } t/k) = 1 - \lambda\, t/k.$$

We can now calculate the probability of a number of arrivals n in the interval t as the product of all possible combinations of k intervals (each of size t/k) taken n at a time, and the probability of n arrivals, and the probability of $(k - n)$ no-arrivals:

$$P(n \text{ arrivals in interval } t) = \binom{k}{n} (\lambda t/k)^n (1-\lambda t/k)^{k-n},$$

which will be recognized as the Binomial distribution.

Since we do not know how large to make k in order for t/k to be sufficiently small, we consider making it arbitrarily large, and look to the limit as it increases:

$$\lim_{k \to \infty} P(n \text{ arrivals in } t) = \frac{(\lambda t)^n}{n!}\, e^{-\lambda t},$$

which is the Poisson distribution.

These distributions are special cases of a class known as Erlang distributions, which correspond to physical systems of considerable scope.

Similar assumptions can be made about the servers. They are:

identical,

never idle while a request is awaiting service,

individually able to serve only one request at a time, and

individually possessed of a negative exponential service-time distribution (reexamine Figure 5-19). (The necessity of this assumption will be questioned in one of the Exercises.)

The last of these assumptions relates to the holding time of the requester, since in telephone systems the *server* is at the mercy of the requester and must serve until released. The negative exponential is thus tied to an assumption about the holding time of the requester:

The probability of a customer terminating a call in the next (sufficiently small) interval of time is independent of how long the call has been in progress, but is proportional to the interval.

More specific assumptions are made (in the interests of solubility) with regard to a requester's reaction to blocking. Extremes include the *Lost Calls Cleared* (LCC) case in which a request that cannot be served leaves the system and does not return, and the *Lost Calls Delayed* (LCD) case, in which a call will wait arbitrarily long for service. These cases essentially bracket reasonable customer behavior patterns (though there are situations for which they are excellent models), the first yielding optimistic blocking probabilities and the second, pessimistic values. A more realistic situation might be visualized as a requester waiting for a time after being blocked, then leaving the system if not served; and for the special case of the average waiting time equal to the average holding time (assuming a negative exponential waiting-time distribution) it is called the *Lost Calls Held* (LCH) case.

An example of Lost Calls Cleared would be the behavior seen by a high-usage trunk group with overflow to another (final) group; an unserved call would leave and not reappear. A Lost Calls Delayed situation occurs in a message switching system, where messages, disembodied from their originator's impatient temperament, passively await service. The Lost Calls Held case is exemplified by a customer's behavior when he does not immediately receive dial tone. The following will undertake the analytical solution of these cases.

It will be assumed that a uniform level of traffic is offered to the system such that a state of *statistical equilibrium* exists. Statistical equilibrium will be interpreted as meaning that the fraction of time that a given number of servers is occupied does not vary with time. Put another way, the rate of departure from the system must equal the rate of arrival for equilibrium to obtain.

Let

$f(i)$ be the fraction of time that i servers (e.g., trunks) are in use,

n be the rate of arrival of calls,

h the average holding time per call, and

a the average number of servers in use.

Then the following expressions can be written:

$$\sum_{i=0}^{\infty} f(i) = 1, \tag{1}$$

$$n \sum_{i=0}^{\infty} f(i) = n, \tag{2}$$

$$\sum_{i=0}^{\infty} nf(i) = n, \tag{3}$$

and

$$\sum_{i=0}^{\infty} if(i) = a. \tag{4}$$

Note that $a = hn$, therefore

$$\sum_{i=1}^{\infty} i/h \ f(i) = a/h = n = \sum_{i=0}^{\infty} nf(i),$$

or

$$\sum_{i=1}^{\infty} i/h \ f(i) = \sum_{i=0}^{\infty} nf(i). \tag{5}$$

While (5) is true in general, the condition of statistical equilibrium places the additional constraint that individual terms of the two series be equal, i.e.,

$$i/h \ f(i) = nf(i-1). \tag{6}$$

This is true because the left-hand side of (6) represents the number of calls per unit time departing from the state of i busy servers (dropping to the state of $i-1$ busy servers), while the right-hand side represents the number of calls per unit time arriving in the system when $i-1$ servers are busy (causing a change to the state of i servers busy).

Thus

$$f(1) = \frac{nh}{1} \ f(0), \tag{7}$$

$$f(2) = \frac{nh}{2} \ f(1) = \frac{(nh)^2}{2} \ f(0), \tag{8}$$

and, in general,

$$f(i) = \frac{(nh)^i \ f(0)}{i!} \ . \tag{9}$$

Substituting (9) into (1), and using $a = nh$,

$$\sum_{i=0}^{\infty} \frac{a^i f(0)}{i!} = 1.$$ (10)

Noting that

$$\sum_{i=0}^{\infty} \frac{a^i}{i!} = e^a,$$ (11)

from (10) we obtain

$$f(0) = e^{-a},$$ (12)

and substituting in (9), we get

$$f(i) = \frac{a^i e^{-a}}{i!}$$ (13)

for the fraction of time i servers are busy.

Lost Calls Held (LCH)

For the Lost Calls Held case, a call entering a system with all its servers, say c, busy, will expend an average holding time either in waiting, or in a combination of waiting time plus service time if a server becomes available before it departs the system. For analysis purposes, we can assume that calls encountering all servers busy will be served by *theoretical servers,* which allows the use of the above expressions.

The number of calls entering the system when all servers are busy is given by

$$\sum_{i=c}^{\infty} n f(i),$$ (14)

which, using (13), is equal to

$$\sum_{i=c}^{\infty} \frac{n a^i e^{-a}}{i!} .$$ (15)

The probability of blocking is the ratio of the number of blocked calls to the number of offered calls per unit time, n, or

$$P\ (c,a) = \sum_{i=c}^{\infty} \frac{a^i e^{-a}}{i!} \quad LCH.$$

Figures 5-29 and 5-30 display the characteristics of the LCH case. It will shortly be appreciated that the properties of this case will be bracketed by the other two cases to be considered.

Lost Calls Cleared (LCC) or Erlang B

Under a lost calls cleared regimen, any calls entering the system with all c servers busy are cleared forever, therefore expressions may be truncated at $i=c$. Expressions (1), (3), (12), and (13) can therefore be altered:

$$\sum_{i=0}^{\infty} f(i) = 1 \qquad \rightarrow \qquad \sum_{i=0}^{c} f(i) = 1 \qquad (16)$$

$$\sum_{i=0}^{\infty} nf(i) = n \qquad \rightarrow \qquad \sum_{i=0}^{c} nf(i) = n \qquad (17)$$

$$f(0) = e^{-a} \qquad \rightarrow \qquad f(0) = \frac{1}{\sum_{i=0}^{c} \frac{a^i}{i!}} \qquad (18)$$

$$f(i) = \frac{a^i e^{-a}}{i!} \qquad \rightarrow \qquad f(j) = \frac{\frac{a^j}{j!}}{\sum_{i=0}^{c} \frac{a^i}{i!}} \qquad (19)$$

The probability of blocking is identically the probability of c servers being in use:

$$f(c) = B(c,a) = \frac{\frac{a^c}{c!}}{\sum_{i=0}^{c} \frac{a^i}{i!}} \quad LCC.$$

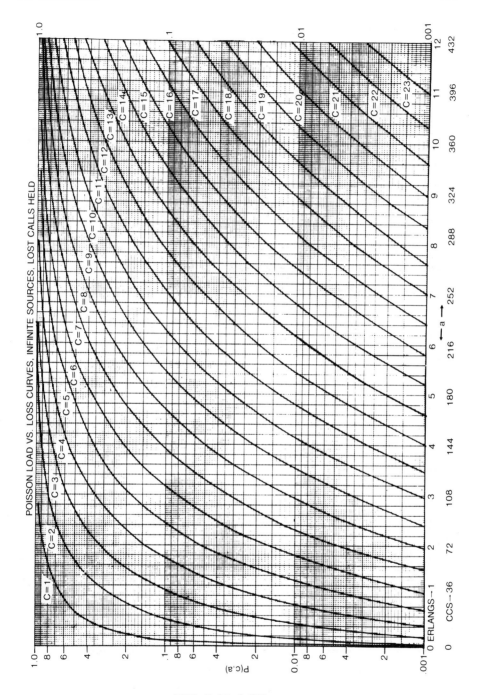

FIG. 5-29. LCH curves.

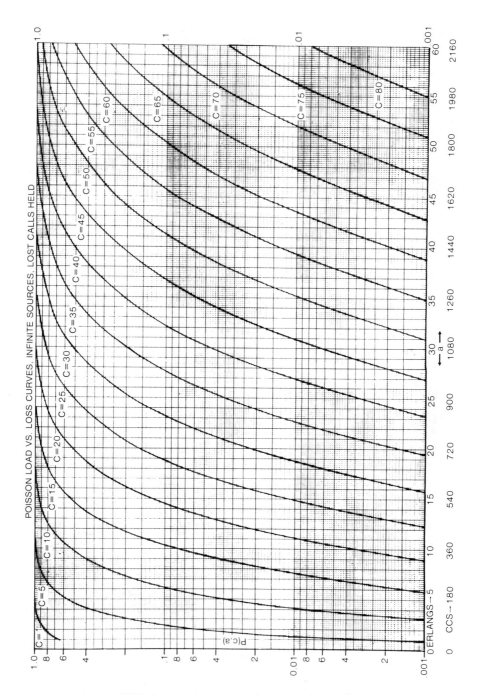

FIG. 5-30. LCH curves (contracted scale).

LCC curves are displayed in Figures 5-31 and 5-32. Note that the curves only asymptotically approach unity blocking. Because blocked sources immediately leave the system, they do not accumulate to increase congestion.

Lost Calls Delayed (LCD) or Erlang C

In a lost calls delayed system, calls remain in the system until served, independent of how long it may take. The terms of equation (6) still apply up to $i = c - 1$:

$$n f(0) = \frac{f(1)}{h}$$

.
.
.

$$n f(i) = \frac{(i+1) f(i+1)}{h}$$

.
.
.

$$n f(c-1) = \frac{c f(c)}{h} \quad .$$

But for i ⩾ c, the terms change:

$$n f(c) = \frac{c f(c+1)}{h}$$

.
.
.

$$n f(i) = \frac{c f(i+1)}{h} \quad .$$

The change reflects the fact that only c servers contribute to consuming the holding time of a call. Summing up the terms for $i > c$,

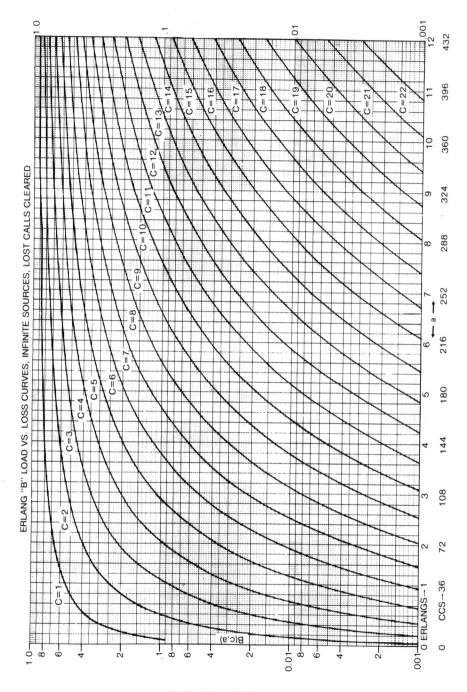

FIG. 5-31. LCC curves.

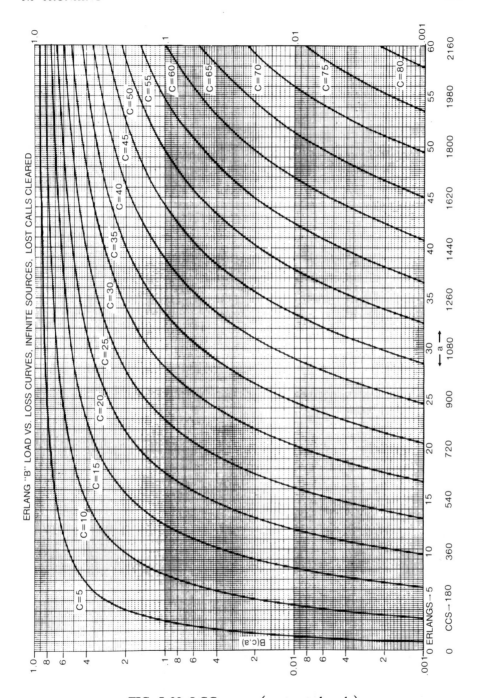

FIG. 5-32. LCC curves (contracted scale).

$$f(c+1) + f(c+2) + \cdots = f(c) \left[\frac{nh}{c} + \frac{(nh)^2}{(c)^2} + \cdots \right]$$

$$= f(c) \frac{nh}{c-nh} = f(c) \frac{a}{c-a} \quad . \qquad (20)$$

Now considering again equation (1),

$$\sum_{i=0}^{\infty} f(i) = 1,$$

and splitting it into the terms occurring before and after $i = c$ and using (9), we obtain

$$\sum_{i=0}^{c-1} \frac{a^i}{i!} f(0) + \sum_{i=c}^{\infty} f(i) = 1 \quad .$$

Substituting from (20), we can write

$$\sum_{i=0}^{c-1} \frac{a^i}{i!} f(0) + f(c) \left[1 + \frac{a}{c-a} \right] = 1.$$

Substituting for $f(c)$ using (13),

$$\sum_{i=0}^{c-1} \frac{a^i}{i!} f(0) + \frac{a^c f(0)}{c!} \frac{c}{c-a} = 1,$$

and solving for f(0), we obtain

$$f(0) = \frac{1}{\displaystyle\sum_{i=0}^{c-1} \frac{a^i}{i!} + \frac{a^c}{c!} \frac{c}{c-a}} \quad .$$

The probability of a call being delayed is

$$\sum_{i=c}^{\infty} f(i) = f(c) \frac{c}{c-a}$$

$$= \frac{a^c}{c!} \ \frac{c}{c-a} \ f(0),$$

or

$$C(c,a) = \frac{\dfrac{a^c}{c!} \ \dfrac{c}{c-a}}{\displaystyle\sum_{i=0}^{c-1} \dfrac{a^i}{i!} + \dfrac{a^c}{c!} \ \dfrac{c}{c-a}} \qquad LCD.$$

Figures 5-33 and 5-34 illustrate LCD behavior. Note that the curves are no longer merely asymptotic to unity blocking, but intersect unity at the point where the number of servers equals the traffic expressed in erlangs. This can be appreciated by noting that blocking is unavoidable in such a situation unless the traffic is arbitrarily *smooth* (see the Nonrandom Traffic Section).

Trunks are often engineered on the basis of the LCC assumption, although the other two are applied in specific cases where they are more appropriate. For low blocking, all three models are so close together that the choice is academic (Figure 5-35). Some examples of the use of these formulas are given:

Lost Calls Cleared Example

One hundred and fifty calls with an average length of 200 seconds are offered to a high-usage trunk group during the busy hour. If the group comprises 10 trunks, what grade of service will be provided?

Solution

Since a high-usage trunk group alternate-routes overflow traffic, this is a lost calls cleared situation. The offered load is:

150 calls x 200 seconds = 300 CCS.

Referring to Figure 5-31, we see that for 10 trunks and 300 CCS load, the blocking probability is about 0.15.

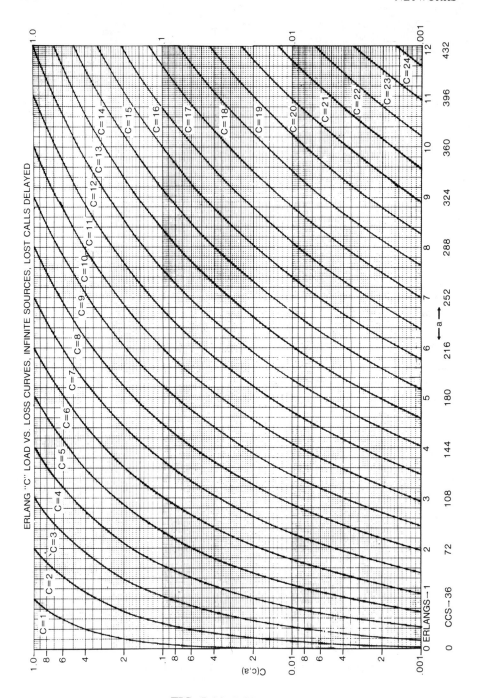

FIG. 5-33. LCD curves.

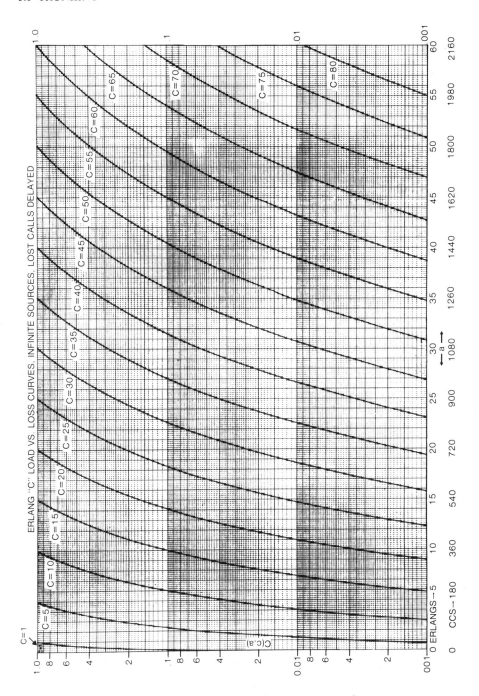

FIG. 5-34. LCD curves (contracted scale).

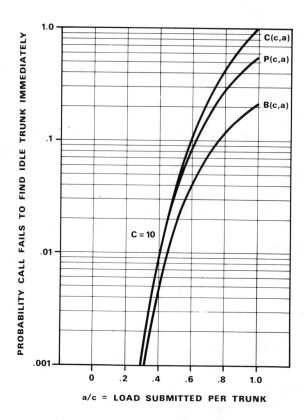

FIG. 5-35. Blocking formula comparison.

Lost Calls Delayed Example

A No. 5 crossbar system has three dial-tone markers providing four calls lost out of 100 grade of service. How many markers must be added to improve the grade of service to 1 out of 100?

Solution

Lost calls are assumed to be delayed in this situation. Referring to Figure 5-33, we find that the offered load is about 28 CCS, which requires four servers for 0.01 blocking. Therefore one marker must be added.

Lost Calls Held Example

A full trunk group is to be engineered for 0.01 blocking under an expected offered load of 360 CCS. How many trunks will be required?

Solution

A full group does not have rerouting capability, so a Poisson behavior is assumed. From Figure 5-29, we find that 19 trunks will be required.

Finite Sources

To this point, it has been assumed that the traffic sources are infinite in the sense that the flux of traffic entering the system is unaffected by the number of sources in the act of being served. (Note that an infinite source assumption also implies that the per-source traffic is vanishing in magnitude in order to have finite total offered traffic.) In a real, and therefore finite, system the entering traffic is diminished when sources are being served. Results under an infinite-source assumption are therefore conservative (indicating a need for more servers than are necessary). For all but systems with a number of sources comparable to the number of servers, however, the infinite source assumption is satisfactory.

An analysis similar to that carried on for infinite sources yields the Engset formula for Lost Calls Cleared, which can be shown to reduce in the limit to the form derived above.

Limited Availability

Some networks do not offer full availability of all outlets to all inlets. Such networks are known as *limited availability* networks.

Typically, in such situations, subgroups of the inlets are allowed access to a fixed subgroup of the outlets. When several subgroups of the inlets are allowed access to a given outlet, the pattern of connection is referred to as a *grading*. Grading patterns are beyond the scope of this book.

Nonrandom Traffic

Thus far only *random* or Poisson traffic has been considered. Traffic distributions actually encountered often deviate from the Poisson. The distinguishing feature between traffic distributions is the variance-to-mean ratio, α. For $\alpha < 1$, the traffic is characterized as *smooth;* for $\alpha > 1$, it is characterized as *rough* or *peaked*. The $\alpha = 1$ case corresponds to Poisson or *random* traffic (see Figure 5-36). In general, distributions with longer tails require more servers than those with shorter tails, for the same mean. Practically encountered αs range from about 0.5 to 2.0.

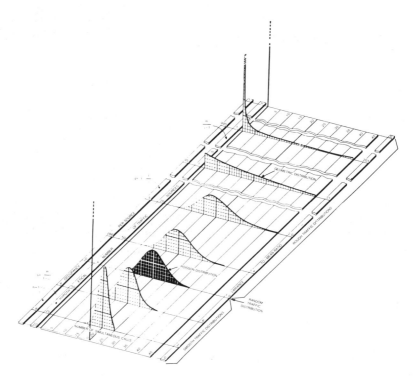

FIG. 5-36. Effect of variance-to-mean ratio.

The traffic offered to overflow facilities is distinctly nonrandom in character, so that they cannot be properly engineered via direct use of the techniques discussed thus far. A method due to Wilkinson and known as the *Equivalent Random Method* may be employed in such cases. This technique makes use of the argument that, for a given offered traffic with mean A and variance V, there corresponds an equivalent Poisson traffic load with mean, A', which, when offered to a number of servers, S, will produce an overflow traffic with characteristics A and V.

If the random load, A' and the value of S can be ascertained, then the number of servers, T necessary to provide the desired grade of service can be calculated. The number of servers necessary to provide the same grade of service to the offered load A, V is then clearly the difference between S and T.

Exact computation of A' and S are relatively difficult, and curves are available that provide adequate precision; Rapp,[6] however, has produced two empirically derived expressions:

$$A' \approx V + 3z(z-1),$$

and

$$S \approx A' \frac{(A+z)}{(A+z-1)} - (1+A),$$

where $z = \dfrac{V}{A}$.

Nonrandom Traffic Example

A number of trunks must be provided to serve an offered traffic load having a mean of 10 erlangs and a variance of 14 erlangs squared, with a 0.01 grade of service.

Solution

$$A' \approx 14 + 3 \times 1.4(1.4-1) = 15.68 \text{ erlangs,}$$

$$S \approx 15.68 \times \frac{(10+1.4)}{(10+1.4-1)} - (1+10) = 6.2 \text{ trunks.}$$

Referring to the Erlang B curves of Figure 5-32, we find that about 24.5 trunks would be required to provide 0.01 blocking; therefore 19 trunks would be required to serve the given, nonrandom load (some 15 trunks would have sufficed if the load were random).

---

6. Y. Rapp, "Planning of Junction Network in a Multi-Exchange Area," *Ericsson Technics,* Vol. 20, No. 1, pp. 77-130.

Delay Engineering

In situations where the LCD assumption is appropriate, it may become necessary to engineer in accordance with a delay criterion. An example of this situation is *dial-tone delay*: when a customer originates during a period of high network congestion, there may be significant delay before dial tone is returned. Engineering the number of receivers in the central office to attain a given delay criterion is therefore desirable. Subject to assumptions about order of service and holding-time behavior, these considerations are amenable to analysis, and curves are available which permit straightforward solution.

Extreme Value Engineering

The use of an *average busy-hour* traffic level in engineering telephone equipment proves inadequate in small offices and loop concentrator systems. In such systems, the high-traffic hour is too variable to permit a meaningful time-consistent busy hour to be identified.

A technique known as *extreme value engineering* is employed in such cases, which, essentially, searches out the busiest hour period in (typically) a week and uses a collection of these measurements to determine a representative traffic load.

Importance of Size

Perhaps the most significant point to be made in considering traffic is the economy of scale. Large groups of servers are more efficient, for the same grade of service, than small groups. Scrutiny of the curves provided may make this statement appear rash, as it will be observed that the statement appears to apply only relative to small groups (Figure 5-37). However, in practice, such small groups are very common, either because the traffic is light between certain entities or because the trunks are quantized into subgroups of limited size.

While large server groups are more efficient, it should be noted that they tend to react poorly to overload because their occupancy is so high.

5.9 PATH HUNT

The path-hunt algorithm employed can have a profound effect upon the performance of a network, though practical considerations may constrain the technique employed to one that is nonoptimum. Thus in electromechanical

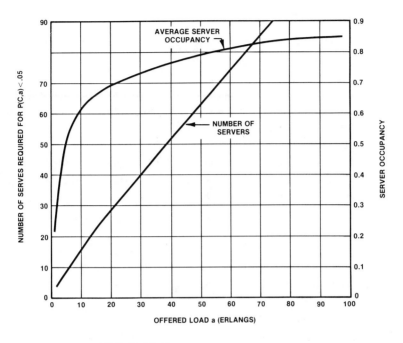

FIG. 5-37. Servers versus occupancy.

networks the need to distribute wear uniformly across the switches may be the overriding requirement.

A pseudorandom choice is suitable in many modern networks but will not in general yield the best possible blocking behavior. The optimum strategy appears often to be one known as *packing,* which consists of first attempting to find a path through the most congested part of the network, and, if necessary, retreating to a less congested portion. While provably better than random choice in simple networks, and demonstrably better (via simulation) in some nontrivial networks, there exists no proof of superiority in general.

It may be speculated that the true optimum algorithm might be one that could be called *precessed packing,* which would pack with respect to a region in the network that would cyclically precess through the network with a period equal to the expected average holding time.

5.10 REARRANGEABLE NETWORKS

It is useful to draw some relatively fine distinctions between three classes of networks:

A network is said to be *nonblocking in the strict sense* if a legal connection attempt cannot be blocked independent of the state of the network.

A network is said to be *nonblocking in the wide sense* if a legal connection cannot be blocked independent of the state of the network so long as a specified algorithm has been followed in setting up all prior connections in the network.

A network is said to be *rearrangeable* if a legal connection cannot be blocked given that the present state of the network can be suspended briefly, and a subset of the connections can be taken down and reestablished using other paths chosen to allow the needed new connection to be made.

The ordering strict-sense nonblocking, wide-sense nonblocking, rearrangeable may be viewed as ranging from the strongest to the weakest requirement to be placed upon a network, and the number of crosspoints needed to meet these requirements follow in the same order.

The problem of mismatch blocking would be eliminated if it were possible to rearrange the existing paths to accommodate a new one. Such an act is impractical in electromechanical networks because their action is too sluggish and would at best produce audible clicks. In a time-division network (e.g., that of No. 4 or No. 5 ESS), however, it becomes possible to seriously entertain rearrangement. However, this would introduce a considerable processing complication, requiring a larger program and more real-time processing, but with both memory and processing capacity dropping in cost, it may become reasonable.

5.11 ANALYSIS OF A PRACTICAL NETWORK UNDER GROWTH

Networks and traffic theory have been discussed thus far as distinct topics, though they are basically inseparable. An example of analysis of some aspects of a small, practical network will serve to illustrate the application of traffic theory and some subtle aspects of intratraffic and partial equipping.

Consider a small switching system employing a folded structure that comprises one or more 2-stage modules as indicated in Figure 5-38.

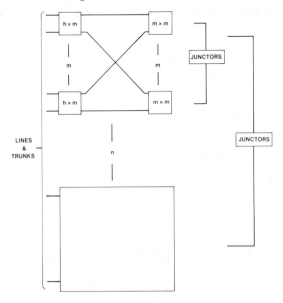

FIG. 5-38. Practical network example.

For n modules, the number of intermodule junctor groups is given by

$$G = \frac{n!}{2(n-2)!} = \frac{n(n-1)}{2},$$

while the number of intramodule junctor groups is, of course, n.

All intrasystem traffic must be carried by the aggregate of inter- and intramodule junctor groups, and the intermodule groups will each carry twice the traffic of the intramodule groups. Therefore, if the fraction of the total intratraffic handled by an intramodule group is x, the following expression must hold:

$$nx + \frac{2n(n-1)x}{2n} = 1,$$

or

$$x = \frac{1}{n + n(n-1)} = \frac{1}{n^2} .$$

The actual intratraffic carried by an intramodule junctor group is given by

$$x \ I \ T/2,$$

where

$I =$ the intraoffice fraction, and

$T =$ the total traffic.

The total traffic T may be expressed as

$$T = n \ T_m$$

where $T_m =$ the per-module traffic.

Therefore, an intramodule junctor group carries

$$\frac{1}{n^2} \ \frac{I \ n \ T_m}{2} = \frac{I \ T_m}{2n} \ \text{ units of intratraffic}[7].$$

The total traffic carried by an intramodule junctor group includes interentity traffic whose magnitude is

$$(1 - I) T_m ,$$

---

7. This result may also be obtained via a more direct but less insight-producing route.

so that the total traffic carried by each such group is

$$(\frac{I}{2n} + 1 - I) \, T_m = (1 - \frac{2n-1}{2n} \, I) \, T_m \ .$$

Thus, for example, for 1 through 4 modules, the intramodule junctor groups must carry

$$n = 1 \quad : \quad (1 - 1/2 \, I) \, T_m \ ,$$

$$n = 2 \quad : \quad (1 - 3/4 \, I) \, T_m \ ,$$

$$n = 3 \quad : \quad (1 - 5/6 \, I) \, T_m \ ,$$

$$n = 4 \quad : \quad (1 - 7/8 \, I) \, T_m \ ,$$

and in the limit as n increases,

$$n \to \infty \quad : \quad (1 - I) \, T_m$$

units of traffic.

An additional constraint is placed upon the traffic-handling capability of the network by the intermodule junctors themselves. Such junctor groups carry traffic given by

$$2/n^2 \times I \, T/2 = I \, T/n^2 = I \, n \, T_m/n^2 = I \, T_m/n.$$

For L lines served per module, and T_g the traffic carried by a given size junctor group,

$$\frac{I \, T_m}{n} = \frac{ICL}{36n} = T_g,$$

where C is the per-line traffic in CCS.

Then

$$C = \frac{36\,T_g}{L\,I}\,n \quad.$$

Thus the traffic limitation imposed by the intermodule junctor capacity as a function of the number of modules is a straight line through the origin with a slope of

$$\frac{36\,T_g}{L\,I} \quad.$$

By way of example, for 440 lines per module and junctor groups of size 32 (which can accommodate 18 erlangs of traffic at 0.001 probability of blocking, lost calls cleared), the slope becomes

$$36 \times \frac{18}{440\,I} = \frac{1.47}{I} \quad.$$

When partial equipping is carried to the point of splitting a module (e.g., in the case of a very small office where only a fraction of the capacity of a full module is required), internal module considerations come into play. Thus, for example, if a module is partially equipped with inlet and junctor switches, the concentration ratio increases (because of the reduction in the number of available links) unless each inlet switch is deloaded. Similarly, unless junctors are rearranged, the number available to a junctor switch is reduced.

5.12 NONSERIES-PARALLEL NETWORKS

The networks examined in Section 5.4 were classifiable as series-parallel. When networks possessing larger numbers of stages are necessary, nonseries-parallel or *spider-web* interconnections are used to improve network performance. Such patterns are somewhat more difficult to analyze, but yield to the technique of partitioning into subgraphs which result from each possible busy-idle condition of a subset of the links.

As an example, consider the network of Figure 5-39, which serves 16 inlets and 16 outlets. Its Lee graph is clearly not series-parallel. Figure 5-40 shows the subgraphs corresponding to the possible A-link conditions. Considering in turn all possible numbers of idle A-links:

$$B(0\ A\text{-links idle}) = 1,$$

$$B(1\ A\text{-link idle}) = [1 - (1 - 4p)(1 - 2p)]^2,$$

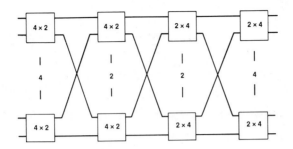

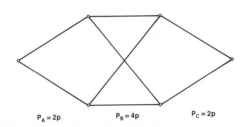

FIG. 5-39. Nonseries-parallel example.

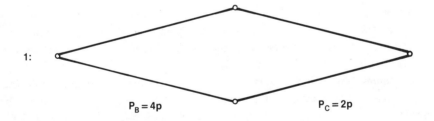

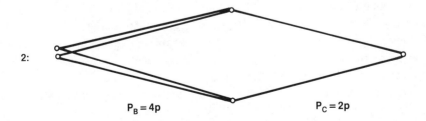

FIG. 5-40. Subgraphs of example network.

$$B(2 \text{ } A\text{-links idle}) = \left[1 - [1 - (4p)^2](1 - 2p)\right]^2.$$

The network blocking probability is then given by

$$P_B = P_A^2 + 2 P_A(1-P_A)[1-(1-4p)(1-2p)]^2$$

$$+ (1-P_A)^2 \left[1-(1-4p)^2](1-2p)\right]^2$$

$$= (2p)^2 + 2(2p)(1-2p)[(1-4p)(1-2p)]^2$$

$$+ (1-2p)^2 \left[1-[1-(4p)^2](1-2p)\right]^2 .$$

As the network size increases, the blocking polynomial becomes increasingly complex, partially explaining the interest in simulation techniques for analysis.

5.13 BROADCAST NETWORKS

Traditionally, networks for voice traffic have been carefully designed to minimize the chances for inadvertent multiple connections (in some, such connections are purposely made impossible). Conference calls were set up using service circuits external to the network fabric.

In some applications, however, there is interest in making *broadcast* connections between some number of inlets and up to all of the outlets. Such a capability has a number of implications upon the design and performance of the network.

Suppose, for example, that the network were intended to be used to switch video signals. Customers are accustomed to being able to switch to any television channel at will, probably requiring a nonblocking switch. Further, customers will have a strong tendency to change channels on the hour and half-hour, requiring rapid response from both the intelligence of the system and the mechanism controlling the network.

Generally speaking, a broadcast network must be more richly endowed than a point-to-point network because the link usage is heavily correlated.

5.14 ANALYSIS OF TIME-SPACE NETWORKS

A time-division or time-space-division network may also be analyzed using the techniques discussed. The Lee graph is determined by visualizing all paths available in a given time slot in parallel with all paths available in other possible time slots. In effect, in each time slot, a unique physical network is available for completing paths.

EXERCISES

1. Use perfect induction to prove that the Clos network of Figure 5-8 is non-blocking.

2. Design a three-stage, $n=4$ Clos network with eight inlets and eight outlets; draw the network with the crosspoints and interconnections displayed. How does the crosspoint count compare with the $n = 2$ case discussed in Section 5.2?

3. Design two three-stage networks and compare their crosspoint counts and probabilities of blocking, assuming the inlet terminal occupancy p is 0.1 and

   a. $N = 8$; $n/k = 1$; $n = 4$.

   b. $N = 8$; $n/k = 2$; $n = 4$.

4. Check the blocking probability of the network of Figure 5-8. Why does it not compute to 0?

5. The No. 5A crossbar system has the network and link graph of Figure 5-41 for a 1960-line version. For a 194.7-erlang offered load, the advertised first trial blocking probability is 0.02. Is this figure accurate?

6. Draw the junctor pattern for the network of Figure 5-41. (Hint: it is unusual.)

7. For a given office, we wish to identify all trunk groups capable of handling at least 10 erlangs of traffic with 0.01 or better blocking probability irrespective of the behavior of the requesting entities under blocking. How would these groups be identified?

8. You are asked to check the engineering of your building's plant facilities. Making reasonable assumptions about the nature of offered traffic,

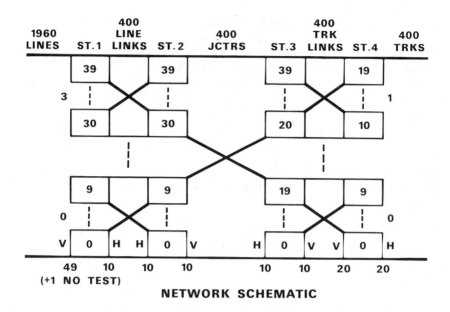

NETWORK SCHEMATIC

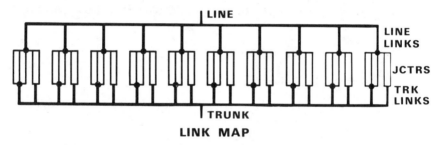

LINK MAP

FIG. 5-41. No. 5A Crossbar network.

calculate the probability of blocking at the washbasins serving your portion of the building. If all such facilities were to be pooled in one large group serving the entire building, how many would be needed, and how does this number compare with the present aggregate?

9. Find the blocking polynomial for the network with the Lee graph given in Figure 5-42. Assume each link has occupancy L.

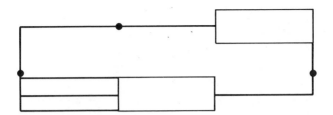

FIG. 5-42. Lee graph for Problem 9.

10. A network serves 1000 lines and 200 trunks. The average per-line traffic is 6 CCS, and intraoffice traffic is 30%. Label the components of an H-diagram representing this network. Label the line and trunk terminals and the junctors of a network diagram with the amount of traffic they carry, assuming the network is folded. Do the same assuming the network is unfolded.

11. Two components of traffic are to be handled by the same trunk group. The first component has a mean of 4 erlangs and a variance of 4 *erlangs*^{2}. The second component has a mean of 6 erlangs and a variance of 10 *erlangs*^{2}. How large must the trunk group be for 0.001 blocking LCH?

12. For the network of Figure 5-43, draw the Lee graph and find an expression for the probability of blocking given that the line occupancy is *p*.

13. There exists a method known as Time Assignment Speech Interpolation (TASI) that is used, for example, to increase the effective capacity of undersea cable. Speech over telephone circuits in one direction occupies the circuit only an average of 35% of the time. TASI "steals" the channel away from a customer when he ceases talking, and assigns it to another, actively talking customer, effectively more than doubling the capacity of the system. A little thought reveals that TASI is a nearly perfect example of one of the traffic types discussed. Consider a representative phrase to be spoken over such a facility; it constitutes a candidate for transmission, but may be *frozen-out* in whole or in part. Find the probability of freeze-out in a TASI system serving 50 (two-way) conversations to be carried on a 25 (two-way) channel cable.

14. Explain why intermodule junctor groups carry twice the traffic of intramodule groups.

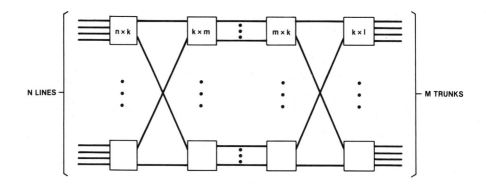

FIG. 5-43. Network for Problem 12.

15. A 4-stage network has $n \times m$ first- and second-stage switches, and $m \times n$ third- and fourth-stage switches. Draw the Lee graph for the network and find an expression for network blocking for a per-line traffic of p.

16. Is the negative exponential service time assumption necessary for the Lost Calls Cleared case?

READING LIST

Mathematical Theory of Connecting Networks and Telephone Traffic, V. E. Benes, Academic Press, 1965. (A very thorough treatment.)

Telecommunications - A Time for Innovation, A. A. Collins and R. D. Pedersen, Merle Collins Foundation, 1973. (Includes a treatment of time-space-time [used, e.g., in No. 4 and 5 ESS] and space-time-space networks.)

Switching Systems, AT&T, 1961. (Somewhat dated.)

A. Feiner and J. G. Kappel, "A Method of Deriving Efficient Switching Network Configurations," *Proc. National Electronics Conf.,* December, 1970. (A good source of insight.)

C. Jacobaeus, "A Study of Congestion in Link Systems," *Ericsson Technics,* Nr. 48, 1950. (Classic paper.)

C. Clos, "A Study of Nonblocking Networks," *BSTJ*, Vol. 32, March, 1953, pp. 406-424. (Classic.)

N. Pippenger, "Complexity Theory," *Scientific American*, Vol. 238, No. 6, pp. 114-124, March 1953.

C. Y. Lee, "Analysis of Switching Networks," *BSTJ*, Vol. 34, November, 1955, pp. 1287-1315. (Classic.)

R. I. Wilkinson, "Theories for Toll Traffic Engineering in the U.S.A.," *BSTJ* Vol. 35 March, 1956, pp. 521-514. (Classic.)

R. V. Lave, "Extreme-Value Engineering Designs for Peak Traffic," *Bell Labs. Record,* Vol. 56, No. 2, February 1978, pp. 38-42.

Telephone Traffic Theory Tables and Charts, Siemens Aktiengesellschaft, 1970. (An essential reference for the serious practitioner, but difficult to obtain.)

CHAPTER 6

STORED-PROGRAM CONTROL

6.1 HISTORY

The first electronic computer (ENIAC) was developed during World War II at the University of Pennsylvania. The machine did not utilize a stored-program structure and was fashioned of vacuum tubes. By modern standards its capacity was tiny, yet it was judged equivalent to some 20,000 people with electromechanical calculators.

In the summer of 1946, a seminar lasting several days was held at the University of Pennsylvania at which ENIAC was described to an international audience. Thoughts on a stored-program machine already under design (EDVAC) were also disseminated. A number of the attendees returned home and promptly began design and construction of their own stored-program machines.

The fundamental realization that instructions could be represented in the same form as data, and modified dynamically in the same way, sparked the invention of the stored-program machine. The strides made in commercial machines since that realization are self-evident.

Though some exploratory work was done, no serious development activity employing an electronic control (wired logic or stored-program) for a telephone switching machine took place in the Bell System until that which led to the Morris, Illinois, field trial in 1960. That machine was a test bed for many switching ideas, the most robust and significant being the stored-program concept, which demonstrated its viability in a telephone environment.

The redesigned and substantially improved version of the Morris system became No. 1 ESS, the first non-PBX commercial stored-program telephone switching machine.

Additional stored-program machines in service in the Bell System include:

No. 101 ESS: A small machine capable of providing PABX service to a multiplicity of groups of customers by remotely controlling a time-division network on the customer's premises.

SP: A satellite processor used in conjunction with No. 1 ESS in high-traffic offices to relieve it of the peripheral communications job.

TSPS: The Traffic Service Position System utilizing the SPC 1A Processor, which is also employed in the 4A Crossbar Electronic Translator System (ETS).

No. 2 ESS: A medium-capacity system for use in moderate-size offices.

No. 3 ESS: A small-capacity system intended for rural or *Community Dial Offices* (CDO); the processor is also employed in the No. 2B ESS system.

No. 4 ESS: A very-high-capacity toll switching system utilizing the 1A Processor, which is also used in the No. 1A ESS system.

No. 5 ESS: A local time-division switching machine using the 3B processor and a multiplicity of microprocessors.

It is worth noting that, with the exception of the 1A Processor, which is essentially upward compatible from the No. 1 ESS, each of the above machines' control processors is unique, with a distinct machine language, and is unable to run programs written in the language of any of the other machines. (The No. 3 Processor, being microprogrammed, could, in theory, emulate other machines, and is, in fact, being used to emulate the No. 2 Processor in the 2B system.)

The investment in programming necessary for No. 1 ESS was of such magnitude that it profoundly influenced the design of the No. 1A Processor, whose order structure was compelled to be (nearly) a superset of that of No. 1 ESS. This approach was taken to guarantee compatibility (in at least the upward direction) between the two machines so that the cost of developing the necessary software would not have to be borne twice.

Meanwhile, the rest of the industry has not been idle; dozens of new systems, many stored program, some experimental, have arisen. Figure 6-1 attempts to depict a selection of the systems; the progression clockwise is: the Netherlands, Sweden, Germany, ITT, France, the United Kingdom, the United States (including some Department of Defense machines) and Japan.

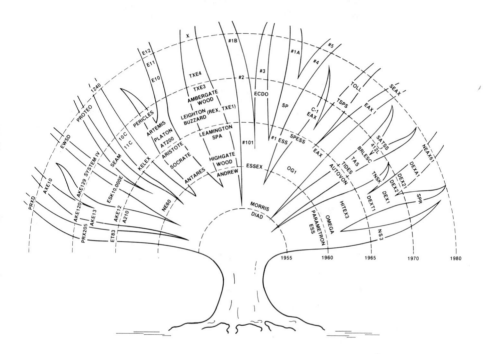

FIG. 6-1. A telephone switching system tree.

6.2 CHARACTERISTICS OF STORED-PROGRAM TELEPHONE PROCESSORS

Someone familiar with the internal design characteristics of commercial computers would be struck more by the similarities than the differences between them and Bell ESS processors (Figure 6-2). The major differences are:

1. A shift in emphasis from arithmetic operations to logical operations (thus, for example, multiply, divide, and floating-point facilities need not be present, while *find rightmost one* is typically provided, as are elaborate masking facilities).

2. A broadened capability for peripheral control and communication (the periphery includes the network, lines, trunks, etc.).

3. Segregation of program and data in different stores in the earlier machines (which has the advantage of increasing processing speed because data can be addressed before the program memory has completed its read cycle; another reason in the earlier ESS machines was the use of Read-Only Memory (ROM) for the program store because of fear of the consequences of the loss of any information in the program [this feature is not unique to telephone processors, but had been somewhat rarely used elsewhere until the relatively recent advent of semiconductor memories]).

4. Elaborate checking circuitry, redundancy, and means for automatic reconfiguration in response to failure.

TELEPHONE	COMPUTER
● WEAK ARITHMETIC	● POWERFUL ARITHMETIC
● POWERFUL DATA MANIPULATION	● WEAKER DATA MANIPULATION
● POWERFUL CHECKING & DIAGNOSTICS	● WEAK CHECKING & DIAGNOSTICS
● USUALLY DUPLICATIVE REDUNDANCY	● ALMOST NEVER DUPLICATED
● BROAD PERIPHERAL COMMUNICATION CAPABILITY	● RELATIVELY NARROW (DIRECT) PERIPHERAL COMMUNICATION CAPABILITY
● PROGRAM & DATA SEGREGATION	● PROGRAM & DATA INTEGRATION

FIG. 6-2. Telephonic vs. commercial processors.

6.3 WHY ASSEMBLY LANGUAGE?

Very soon after the first computers were constructed, it was recognized that it was folly to write programs in binary.[1] Instead, mnemonics began to be employed, and programs (assemblers) were written to convert (assemble) a program written using the mnemonic *assembly* language into machine-intelligible binary. The first assembly languages were relatively primitive, but over the years they have grown quite powerful. The distinguishing features of an assembly language are that it is machine-dependent, and, in general, converts under assembly from one assembly language statement to one binary machine instruction (see Figure 6-3).

Binary: 1 0 1 1 0 1 0 0 1 0 1 0 0 1 1 0

Assembly: Add, 1, 3, 137

Pol: If X > Y Then P = P + K; Else P = P-K

FIG. 6-3. Language levels.

In an environment of considerable skepticism, the first notable Problem-Oriented Language (POL) was developed in the 1950s. This language

---

1. Or decimal (some early machines were decimally oriented).

(FORTRAN) opened a Pandora's box which appears unlikely ever to close. There are now a bewildering variety of POLs; examples of the most widely used (besides FORTRAN) are ALGOL, COBOL, and PL1. The POLs or *high-level languages* are distinguishable from assembly languages in that they are (at least in theory) machine independent, and that the conversion (compilation) to binary machine instructions is, in general, a one-to-many process.

Major advantages of POLs are (Figure 6-4) that they ease the programming job (because the programmer need not know the details of the machine upon which the program will be run, and the program is much shorter in POL form because the statements are more powerful), and the program is transportable to any other machine for which a compiler for the POL has been written (in practice there are some difficulties in effecting the latter result).

	ASSEMBLY LANGUAGE	PROBLEM ORIENTED LANGUAGE
WRITING	SLOW & TEDIOUS	RELATIVELY EASY
UNDERSTANDING	DIFFICULT	RELATIVELY UNDERSTANDABLE
DEBUGGING	SLOW & TEDIOUS	EASIER
OBJECT CODE	SPACE & TIME EFFICIENT	INEFFICIENT IN SPACE & TIME (IMPROVABLE AT EXPENSE OF LONGER COMPILATION)
TRANSPORTABILITY	CANNOT (EXCEPT VIA SIMULATION OR EMULATION OR TO COMPATIBLE MACHINE)	TOTALLY (TO A MACHINE POSSESSING A COMPILER)

FIG. 6-4. Assembly language vs. POL.

Why then were not programs for all ESS machines written in a Switching POL? Why were No. 101, No. 1, and No. 2 ESS programs written in (different) assembly languages? The answer has two facets:

The compiled or object code produced by a compiler in converting POL source code is almost invariably[2] less efficient in space (number of instructions to be stored in memory) and time (real-time consumed in running the program) than an equivalent program written in assembly language by a "good" programmer.

Maintenance programs, which must exercise the internal structure of the machine, must take cognizance of the machine's details. They are, therefore, by definition machine dependent, and cannot be written in a machine-independent language. (Maintenance programs typically account for one-half or more of the program memory space, though only a fraction of them are directly concerned with the control processor.)

The inefficiency in memory space translates to higher cost (many times over because of the many machines produced), and the inefficiency in time translates to less call-handling capacity and earlier machine exhaustion, which also has a dollar value. Thus it was decided that the early ESS programs would be written in an assembly language, which, however, has certain powerful facilities (e.g., macro capability) that were not characteristic of conventional assemblers at the time.

No. 5 ESS, due to advances in memory and processor technology that have reduced the cost and increased the performance of memories and processors, utilizes a higher-level language: *C*.

The International Standards Committee CCITT has recommended a switching-oriented language called *CHILL* which is being adopted in many manufacturers' switching machines.

6.4 ARCHITECTURES

By the architecture of a control is meant its arrangement and number of internal registers, its bus structure, its arithmetic and logic capability, its internal control philosophy (wired or microprogrammed, synchronous or asynchronous, etc.), its representation for negative numbers, etc.

---

2. Though there are some glimmerings of hope.

The architecture of a processor can have a profound effect upon the ability of the machine to efficiently run programs of a particular kind (and a corresponding effect upon the programmer's ability to write efficient code). Thus, for example, a highly mathematical program would be awkward to write for an existing ESS (and would run poorly), yet a logical manipulation program can be written with relative ease and would run more efficiently than on many commercial general-purpose computers.

Figure 6-5 shows the relative architectural features of the processors driving current ESSs in the field.

SYSTEM	NO.101	NO.1	SPC 1A	NO.2	NO.3	1A
ADDER	N	Y	Y	Y	Y	Y
INDEXING	N	Y	Y	N	Y	Y
GEN. REGS.	N	N	N	N	Y	Y
STACK	N	N	N	Y	Y	Y
DATA BUSES	1	2	3	1	1	2
INSTS./WORD	1, 2	1	1	1, 2	1, ½	1, 2
MAIN STORAGE	PS: PMT CS: FS	PS: PMT CS: FS	PBT	PS: PMT CS: FS	SC	PS: FC/SC CS: FC/SC
INST. CYCLE	4,8 μs	5.5 μs	7 μs	3-6 μs	1.1-23μs	0.7, 1.4 μs
CIRCUITS	RTL, 100ns	DTL, 35ns	DTL, 35ns	RTL, 35ns	T^{2}L, 7ns	T^{2}L, 7ns
CONTROL	SEQ.	SEQ.	SEQ.	SEQ.	μ-PROG.	SEQ.

FIG. 6-5. ESS processor characteristics.

The increasing confidence in memories can be seen in the progression from the read-only program store of No. 1 ESS with Hamming code checking, through the magnetic core store initially intended for No. 1A ESS with extended parity, to the semiconductor memory now applied in that application and in the 3B processor.

It is interesting to note that the No. 101 ESS had no adder, a facility found in virtually all computers since ENIAC. This fact illustrates that the telephone processing job does not innately require addition (i.e., computation) means, they are merely a convenience in most stored-program telephone processors.

It is worth noting also that there are no index registers[3] in No. 2 ESS; this lack is compensated by means for counting loops and logically (as opposed to arithmetically) modifying an address.

ESSs No. 101, 1, 2, SPC 1A, and SP processors are classical single-accumulator machines, while Nos. 1A and 3 ESS, and the 3B20 have adopted a multiple-accumulator or *general register* approach. All except the No. 3 ESS and 3B20 processors have classical, sequential circuit controls (in the telephone office, the entire processor is referred to as the *control*; here the term is applied to the controlling portion of the processor, as opposed to the controlled portion [e.g., the arithmetic and logic manipulation part]); the No. 3 ESS and 3B20 processors utilize a microprogrammed control.

The major difference between a sequential circuit and a microprogrammed control (Figure 6-6) is that the latter permits relatively simple alteration of the control (by changing the contents of the control memory); such a change alters the machine's instruction repertoire and permits at will (sometimes with some strain) the *emulation* of another machine. When one machine can emulate another, it can run programs written for the other machine. If the control memory is electronically alterable, a machine can make a "Jekyl-Hyde" transformation almost instantaneously.

6.5 OTHER POSSIBLE STORED-PROGRAM MACHINE ARCHITECTURES AND FEATURES

The machines discussed thus far are basic uniprocessor systems, with the exception of No. 1 ESS which has an optional signal processor, and No. 5 ESS which can be viewed as a distributed multiprocessor. Further, the controls fall in the category of traditional Von Neumann machines. Other processor architectures and philosophies are possible, some of which have been applied in telephonic applications.

---

3. Index registers increase the space efficiency of programs executing repetitive functions or loops; they also relieve the programmer of the need to consciously perform dynamic instruction modifications via arithmetic and logical manipulation.

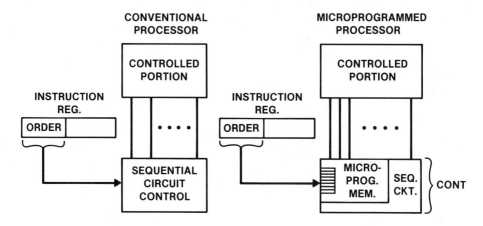

FIG. 6-6. Microprogrammed vs. conventional processors.

Multiprocessors

Where increased ultimate capacity or smooth capacity increase with growth are sought, multiprocessor arrangements become of interest. Crossbar systems employ markers in this spirit, and a No. 1 ESS - SP combination may be viewed as a limiting case.

In addition to increased maximum capacity, and smooth growth, graceful failure is possible with multiprocessor architectures. The smooth growth results from the ability to add additional processors as additional capacity is required. Graceful failure results if each processing element is sufficiently self-checking such that a failure can be quickly recognized, and the crippled element can automatically shut down, reducing the switching machine's capacity by only the fraction thereof represented by that machine.

Potential disadvantages (or reduced advantages) include decreasing per-unit efficiency (caused by interprocessor interference), increasing costs of interconnection means with increased numbers of units, and administration problems.

The two most common modes of operation are direct and functional multiprocessing. Direct multiprocessing allows each processing element to perform all control functions, but economics demand shared storage, which makes interference possible and requires a massive (and expensive) interconnection arrangement. Functional multiprocessing delegates a limited set of tasks to each processor, which then possesses private memory. However, operation under failure conditions becomes a problem, as does growth administration, and a form of stranded processing capacity may result if the tasks are not equitably distributed among the processors.

Language-Oriented Machines

One way in which the inefficiencies of a high-level language might be overcome is through the use of a processor architecture specifically geared toward executing either the high-level language or an intermediate language derived therefrom.

If the process of compilation is analyzed, it conventionally will be seen to consist of two steps, the first converting the high-level code into an intermediate language, the second generating machine code from the intermediate language. The first step (parsing) is relatively straight-forward, the second can be quite complex, especially if efficient code is desired. The second step can be identified as the major culprit in producing inefficiencies, so that its elimination would be highly desirable.

The intermediate language is typically a form of *Reverse Polish* or *Post-Fix* notation, which forms a tree or *stack* equivalent to the parenthesis-free representation of the code. Such a notational language can be directly executed upon a processor designed to be *stack-oriented*. (Parenthesis-using representation is referred to as *In-Fix* notation.)

Figure 6-7 depicts the manner in which a stack-oriented processor executes a program to compute the value of a simple algebraic expression. The processor may be viewed as consisting of an arithmetic and logic unit which performs an operation (monadic [e.g., SQUARE ROOT] or dyadic [e.g., ADD]) upon the top or top two (respectively) elements in a Last-In, First-Out (LIFO) stack. Operands, as they stream in from memory, are recognized as such and pushed down into the stack. Intermediate results are also pushed down in the stack and reappear automatically when required for an operation.

A valuable advantage in speed can be gained in the handling of subroutines and interrupts if the stack is faster than main memory. Not only can subroutine return addresses propagate down in the stack while the subroutine is in control, but the *environment* of the calling routine will propagate down as well, reappearing as the subroutine is completed. Similarly, interrupts can be

Conventional (Infix): a (bc + d)

Reverse Polish (Postfix): abcxd + x

Stack Implementation

Input String:	**Time** →						
	a	b	c	x	d	+	x
Stack Depth ↓	a	b	c	bc	d	bc + d	a (bc + d)
		a	b	a	bc	a	
			a		a		

FIG. 6-7. Reverse Polish notation and stacks.

serviced smoothly and efficiently. (In conventional systems, both subroutine jumps and interrupts require significant overhead because of the need to save, then restore, most registers.)

Pipe-Lining and Cache Stores

"Top-of-the-line" commercial computers make productive use of features such as *pipe-lining* and *cache* memories. These take advantage of the fact that the variety of computation typical of commercial computing has substantial components of straight-line arithmetic code and data base massaging.

The arithmetic unit in a pipe-lined machine has its elements partitioned into portions separated by buffer registers, so that one or more additional operands may enter and be operated upon while a first set is completing its progress. Thus, after a small initial time penalty, the rate of completion of a string of arithmetic operations can be quite high.

The effective memory-access time of a machine can be cut down significantly if a small, additional, very-high-speed store is used as a buffer to hold contiguous blocks of information (or pages). When a memory access is made, the block of information containing the accessed location is written into the cache store, and successive references to adjacent memory locations are automatically made from that store.

The nature of code written for telephonic applications regrettably does not fall in a category that can take maximum advantage of either of these techniques, because very few instructions are typically executed before a branch becomes necessary. However, when brought to bear upon telephonic code, the compilation process can improve the utility of these architectural features.

6.6 AN EXAMPLE OF NO. 1 ESS CODE

The nature of coding for an ESS machine is difficult to convey in a brief exposure. Most technical people today have had experience in programming some machine or in some language. Typical programming experience does not, however, prepare an individual for programming a machine such as No. 1 ESS. Its instruction repertoire is remarkably rich and varied, to be sure, but familiarity with the repertoire is a relatively feeble credential. The program structure itself is the more formidable learning task. Typical programmers are not significantly productive for perhaps a year, and the average ESS programmer productivity is from 500 to 700 written and debugged assembly-language statements per year.

There are several reasons for this seemingly low output. Programming may be viewed as consisting of three operations: formulating a workable algorithm; coding it; debugging the resulting program. The first is made difficult by the complexity of the switching operation and the many interfaces with other programs (and programmers) that are typically necessary. The second is complex because of the extraordinarily powerful repertoire and the need to be cognizant of the hardware structure of the machine. The last strives for an unreachable goal because there is no known way to totally eliminate bugs from a nontrivial program suspended like a plum in a huge pudding of other programs (the best that can be done is to grind them down to a negligibly small number, and to apply defensive programming techniques).

The following example is a genuine sample of No. 1 ESS code, chosen to illustrate some aspects of "typical" ESS programming within the constraints of brevity and understandability without much background. The program performs a recognizable and very simple function: seizing a block of memory called an *originating register* (OR), which, like its namesake in electromechanical equipment, stores dialed digits and is also used for call supervision and outpulsing. The program is a subroutine called by *client* call processing programs.

Idle ORs (and other similar equipment) are kept in a 1-way linked list administered by a 2-word *head cell* containing the addresses of the first and last OR in the list. When an OR is needed, the first in the list is removed (and the links repaired behind it), and when it is no longer needed, it is returned to the list. One more OR is provided than the sum of the receivers and transmitters, so that, since an OR is always associated with one or the other, there should never be fewer than one OR in the linked list (if the list is found empty, it is an error condition, and is tested for by the program).

The integrity of the head cell is critical, for if its contents are "garbaged" by some external influence, an attempt to utilize them will cause an interrupt without correcting them. Each subsequent attempt to seize an OR would have the same effect. As a defense against this possibility, the head cell is zeroed before its prior contents are used, so that the next attempt to seize an OR (assuming the prior contents had been unreliable) would find a zeroed head cell, and the subsequently-called audit programs would correct the condition.

Following is the actual listing and programmer's comments:

1.	MX	Y4HOR	HDCL+0 POINTS TO MOST IDLE REG
2.	TCLE	SAHELP	0=NO IDLE REGS
3.	XM	S6SNAP+5	UPDATE WITH ADRS OF SEIZED REG
4.	EZEM	Y4HOR	PREVENT MULTIPLE INTERRUPTS IF GARBAGED
5.	MB	1,X	Q-WORD OR SEIZED REG POINTS TO NEXT REG
6.	BM	Y4HOR	LINK HDCL+0 TO NEXT REG
7.	EZEM	17,X	
8.	ENTJ	1,J	SUCCESS RETURN TO CLIENT
9.	LM	4YIOR*1(Y4RI)+ 2(Y4PT),X,PS	SET Y4RI=RI, Y4PT=PT

Y4HOR is the symbolic address of the head cell. SAHELP is the symbolic address of an error routine. S6SNAP+5 is the symbolic address of the call register currently active. 4YIOR is the register ID (RI). PT stands for program tag.

A flowchart is hardly necessary since there are only two exits (to SAHELP and back to the client routine, whose address is in the J-register).

Statement 1 — Loads the first word of the head cell into register X.

Statement 2 — Tests the loaded word for 0 or negative contents and branches to SAHELP, which will print an error message and flag the system audits on.

Statement 3 — Places the seized register address in S6SNAP+5.

Statement 4 — Zeroes the first word of the head cell as a precaution against its unreliability.

Statement 5 — Loads register B with the address of the next OR in the list (from within the seized OR).

Statement 6 — Stores the above address in the first word of the head cell, repairing the linked list after the excision of the seized OR.

Statement 7 — Zeroes the seventeenth word of the seized OR to clear out data items from the last call it handled.

Statement 8 — Branches back to the client program, but not until after Statement 9 has been executed (a machine cycle is saved by this peculiar-seeming juxtaposition).

Statement 9 — Initializes the OR state for its impending tasks.

EXERCISES

1. How many instructions are *really* necessary in the repertoire of a stored-program processor in order for it to be able to function as computer? (Hint: Think philosophically.)

2. Convert $a^2 + \dfrac{b(c-d)}{e-gd}$ to Postfix notation.

3. Convert $abcd+\times ef+g-\times-$ to Infix notation.

4. Invent another way to prevent the problem that the No. 1 ESS program sample thwarts by zeroing the head cell temporarily.

5. Can you think of any reasons why the choice of negative number

representation (e.g., 2's complement vs. 1's complement) might have profound effects upon the hardware and software of a telephone processing computer?

READING LIST

Computer History

Computer Structures, Readings and Examples, C. G. Bell and A. Newell, McGraw-Hill, 1971. (A fine historical treatment, including reprints of many classic papers.)

ESS Processors

"No. 1 Electronic Switching System," *BSTJ*, Vol. 43, September, 1964, Special Issue in two parts; also issued as *Monograph 4853*

"No. 2 ESS," *BSTJ*, Vol. 48, No. 8, October 1968, Special Issue.

"TSPS No. 1," *BSTJ*, Vol. 49, No. 10, pp. 2417-2732, December, 1970, Special Issue.

"The 1A Processor," *BSTJ*, Vol. 56 No. 2, February, 1977, Special Issue.

H. E. Vaughn, "An Introduction to No. 4 ESS," IEEE *International Switching System Symposium Record*, June, 1972.

R. D. Royer et al., "No. 5 ESS - System Architecture," *ISS 81*, Vol. 3, pp. 1-6.

Programming Languages

Programming Languages: History and Fundamentals, J. E. Sammet, Prentice Hall, 1969.

Microprogramming

Microprogramming: Principles and Practices, S. S. Husson, Prentice Hall, 1970.

Stack-Oriented Machines

E. A. Hauck and B. A. Dent, "Burroughs B6500/B7500 Stack Mechanism," *Proceedings SJCC*, Vol. 32, 1968, pp. 245-251.

CHAPTER 7

WORLD
SYSTEMS

7.1 GENERALITIES

Switching systems worldwide are increasingly becoming stored-program oriented. Electromechanical and electronic wired logic controllers lack the flexibility, maintainability, and (ultimately) the economy of stored-program processors. While similar to commercial computers, most stored-program telecommunications processors have features unique either in nature or degree of application. There also exists a remarkable variety of overall architectural features that are useful in telecommunications applications, and a similar variety of administrations and manufacturers willing to espouse them.

Some manufacturers have chosen to apply controllers that are also marketed as commercial computers. Such applications were often compromises; either they were procured from a computer manufacturer and inefficiently forced into the mold of telephone processing or were manufactured within and suited one or the other of the two applications suboptimally. This effect is diminishing with time as more flexible processor architectures evolve.

Though many processors in the field were strongly influenced by the conservative philosophies of the earliest versions, some relatively unique if not adventurous machines have been deployed.

The advent of the Integrated Services Digital Network (ISDN) philosophy has greatly influenced the architectures of new machines worldwide. An ISDN provides not only for voice switching and transport, but also data communication, interactive videotex, and other services. A switching system in such a network must provide a more generalized and digitally oriented internal switching capability, and the newest systems are designed accordingly.

7.2 DETAILS

The following are brief descriptions of non-Bell switching machines of note around the world. It will become abundantly clear to the reader that the philosophy of stored-program control has been embraced by most telephone administrations and telephone equipment manufacturers in the world; it is

201

apparently an idea whose time has come. Beyond this generalization, it is more difficult to go, because the modes and manners of implementation vary considerably. However, the increasing popularity of certain architectural aspects is evident; digital time-division switching, multiprocessing, load sharing, $N + 1$ redundancy, microprogramming, stack capability, microprocessor usage, use of modern protocols for packetized internal communication, etc.

United States

General Telephone and Electronics

General Telephone and Electronics has produced a number of electronic control switching systems: the C1 EAX, and Nos. 1, 2, 3 and 5 EAX. The C1 EAX was designed in Canada at Automatic Electric Ltd. It is unique in possessing neither a call store (nor its equivalent) nor an interrupt mechanism. The processors are paired, but one is used in a standby mode. The memory is comprised of a mechanically "written" Dimond ring structure. The latter four machines were designed at Automatic Electric Company in the United States.

All but No. 1 EAX are stored-program-controlled machines; No. 1 EAX is a hybrid, with a substantial component of hard-wired electronic control augmented with a stored-program machine. The philosophies of SPESS, an experimental system designed in the mid-1960s, were applied in the processors of Nos. 2 and 3 (as well as the TSPS System), but No. 3 is a multiprocessor system utilizing up to four paired processors. Only maintenance interrupt capability is provided. A 2B processor, utilizing a triplication and voting philosophy, was made available to upgrade the capacity of No. 2 machines.

No. 3 is a digital switcher intended for the domestic toll market (an international version is called the No. 3I). It utilizes a network structure that is the inverse of the Bell No. 4 and 5 ESS machines: it is essentially an STS switch rather than a TST one.

The No. 5 EAX is a digital switcher for the domestic local exchange market. Its network utilizes a TST structure, and its control employs distributed, duplicated, matched Intel 8086 microprocessors. Remote Switch Units are provided for application in appropriate areas, and digital T1 span lines are used to link the remote units with the Base Unit. The switch is of the TST type, and the entire switch ensemble is in the Base Unit.

North Electric

In 1970, North Electric produced the NX-1E system, which utilizes up to four processor pairs in an autonomous, load-sharing arrangement. Each pair comprises an active and a standby, nonmatching processor.

In 1975, it began deployment of the ETS-4 toll machine, which is an "Americanized" version of the Ericsson AKE13 system, using integrated circuits and a mechanically latching "codebar" switch.

The company's first (local) Digital Switching System (DSS-1) was installed in late 1978. It utilizes a three-level hierarchy of control, the first two employing microprocessors; a load-sharing mode of operation is used with individual processor duplication.

In the Fall of 1977, ITT purchased North Electric's manufacturing and telecommunications-related divisions, which are now called ITT North Electric. The DSS-1 was renamed the 1210 as part of ITT's line of System 12 digital switching systems.

Stromberg-Carlson

Stromberg-Carlson is producing the System Century family of digital central offices. A specialized "telephony preprocessor" is juxtaposed between the periphery and a member of the Digital Equipment Corporation's PDP-11 series of machines. These pairs are duplicated and synchronized with their duplicates.

An additional PDP-11 processor is provided for maintenance purposes.

Over one-half million lines (over 200 systems) had been sold by 1982.

Plessey Telecommunications now owns the Stromberg-Carlson main exchange business.

Vidar

Vidar, formerly a division of Thompson Ramo Wooldridge, produced a digital switching system first introduced in 1976, which is now called the ITS. The ITS uses a STS, strictly nonblocking switching topology in the base unit, with a concentrating (blocking) switch used for class 5 applications. The control is microprocessor-based. More than 250 systems have been installed.

Canada

Northern Telecom (formerly Northern Electric)

Northern Telecom's licensee association (now terminated) with the Bell System influenced the structure of the processor controlling their SP-1 System. Thus a synchronized processor pair and separate ferrite sheet temporary memory and piggy-back[1] twistor memory are utilized. Bell System philosophies are evident elsewhere in the processor as well, but are not universally applied.

The company introduced its CDO version (the DMS-10) of the DMS family of digital switching machines in 1977, and its large local switch (the DMS-100) and toll switch (the DMS-200) in 1979. Microprocessors (peripheral processors) are associated with each peripheral module, while call processing is controlled by a stack-oriented, central processor pair run in synchronized-matched mode.

While the DMS-10 (more than 500,000 lines and 300 systems by 1982) uses a TST network configuration, the 100- (more than 500,000 lines and 100 systems by 1982) and 200-machines use a four-stage time division network.

United Kingdom

The British Post Office has, over the years, encouraged field trials of systems with electronic control, though principally wired logic (ANDREW, HIGHGATE WOOD, LEAMINGTON SPA, LEIGHTON BUZZARD [REX],[2] and AMBERGATE WOOD[3]). Deployed systems include the TXE 1, 2, 3 (wired logic), and TXE 4 (action translator).

More recently, System X has received support and was expected to be controlled by elements resembling the GEC Mark II and Plessey System 250, both of which are multiprocessor, stored-program-controlled machines. The actual design utilizes a 2900 series bit-sliced structure coupled to a 2911 microprogram sequencer. The processor architecture takes advantage of some pipe-lining techniques. Four processors are arrayed upon common buses and

---

1. An electrically alterable version of the twistor memory.
2. Field trial for TXE 1.
3. Field trial for TXE 2.

function in a load-sharing manner.

France

The French have supported field trials for some years with the Ancient Greek Series: SOCRATE, PLATON, ARTEMIS, ARISTOTE, and PERICLES, all but the second of which were stored-program controlled.

The French administration, the PTT, has chosen for deployment a system called El, which includes three machine classes of differing capacity, E10, E11, and E12, the last two of which are stored-program controlled and employ a load-sharing philosophy (E11) or functional multiprocessing (E12).

A newer control arrangement for E10 intended to double its call handling capacity employs a stored-program processor that is microprogrammed and used on a functional multiprocessing basis. The resultant system is called the E10B.

CIT-ALCATEL, manufacturers of the E10 and E11, designed the J2000 processor for telecommunications use, which utilizes up to 31 (one standby for every seven) 16-bit "frontal miniprocessors" for network control, and two 32-bit "central miniprocessors" for call processing. The frontal processors are used in five functional classes.

The same company has recently designed the E10.S system (Americanized version known as the TSS.5). The system utilizes microprocessors exclusively, in both distributed and centralized configurations. The network utilizes an STS philosophy.

West Germany

After developing several experimental systems culminating in the commercial ESK 10,000 E System, which utilized a relatively simple stored-program controller for large offices, Siemens produced a prototype of a new approach designated System IV in 1967. This system, manufactured for the Deutsche Bundespost by several German firms under the generic name EWS, embraces stored-program control fully, utilizing a synchronized processor pair (the SSP 103 processor). The EWSA does analog speech-path switching, while the EWSD does digital switching. The EWSD system utilizes Intel 8086 microprocessors for small systems, and for very large systems, a microprocessor-based multiprocessor called the SSP 113 D.

Siemens has also developed the ESK Crosspoint 44 System, which makes use of functionally employed, duplicated SSP302 microcomputers, and a pair of

SSP201 minicomputers in centralized control, with the possibility of adding another, unduplicated "feature" processor where needed.

Sweden

The Swedish Telecommunications Administration field trialled the A-210 stored-program machine in 1970, and L. M. Ericsson developed the AKE 11, 12, and 13 stored-program systems, which are local, local and transit (tandem), and toll and transit, respectively. Installed AKE 13s utilize the APZ 130 processor, but a more powerful, though compatible processor has become available: the APZ 150.

Ericsson's ARE system is basically intended for control modernization of crossbar, using stored-program techniques.

The Ellemtel, (Ericsson, Swedish Telecommunications Administration) AXE System employs a processor complex denoted the APZ 210, which consists of a large, duplicated processor and a multiplicity of small, "regional" processors (also duplicated). Recently, the AXE 10 system has been augmented with the introduction of a digital subscriber stage, which includes a concentrating time switch and microprocessor-controlled line circuits.

The Netherlands

Phillips manufactured the ETS No. 3 System for trial purposes in 1967; its control was stored program, and a ferrite-core memory was employed.

The first PRX System was cutover in 1972. The philosophy adopted for this machine was to provide one processor size (the TCP 18) for small applications and a larger processor (the TCP 36) in a multiprocessor configuration for large systems. The TCP 36 possesses a subroutine return address stack (64 words deep), a general use stack, a cyclic buffer structure, and DMA capability. A new system, the PRX/D is a digital switcher with a new set of control processors.

Italy

A wired-logic-control system (SEAM) was installed in Rome on an experimental basis in 1965. Present effort is being devoted to the PROTEO System, whose first generation began production in the late 1970s. A duplicate and match philosophy is applied in the control, the 24-bit processor is microprogrammable, and a subroutine return address stack structure is provided. A second generation that is fully digital is undergoing development, and deployment has begun.

Japan

The Japanese were among the first to begin investigation of electronically controlled switching machines, with a parametron[4] logic machine before 1960. The OMEGA, Hitex 3, and DEX machines (DEX T1, 1, 2, 21, A1, A11) followed. The commercial versions of the DEX 2 and DEX A11 are called the D10 and D20, respectively. The most recent machine type is the NEAX 61, a digital switcher usable for both local and toll applications. Its network uses a TST topology.

These machines have typically been designed and manufactured jointly by two or more of Nippon Electric, Oki, Hitachi, or Fujitsu in association with the Nippon Telephone and Telegraph administration.

International Telephone and Telegraph

It is convenient to consider the products of this multinational corporation as a group rather than to associate them with the many telephone administrations that have adopted them.

The Metaconta 11 and 10 machines ("metabar" and reed, respectively) employ ITT 1600, 3200, or 3202 processors, depending upon capacity requirements. These are employed in a load-sharing manner, with $N + 1$ redundancy. The TCS5 employs the ITT 1650 or 1652 processors in a load-sharing configuration; the ITT 1650 was manufactured for ITT by General Automation, the 1652 is a cost-reduced version manufactured by ITT.

ITT acquired the telecommunications segment of North Electric in 1977. The DSS-1 was renamed the ITT 1210 and is being offered for North American sales; it is the first member of the ITT System 12 family of digital exchanges.

The ITT 1240 is the newest and most unique of the ITT exchanges. It uses Intel 8086 microprocessors in a completely distributed multiprocessor

---

4. A parametron is a device capable of being excited in one of two phases by a small perturbing signal. It is powered by a sinusoidal signal running at twice the parametron frequency.

configuration that requires no central processor. The speech path set up is philosophically similar to SXS in that a set of message packets is dispatched into the network, and each crosspoint is equipped with the logic necessary to decode its portion of the header and decide which outlet to choose toward the next stage of switching. The path is thus set up without requiring any a priori knowledge of the busy-idle status of any of the links.

7.3 FUTURES

Future telecommunications system design will be substantially affected by the advent of the microprocessor, which is less likely to be custom-designed for telecommunications applications. Arrays of such machines are likely to be employed in both classical and novel configurations. Functions peripheral to the central processor (if such an entity is identifiable) will tend to be relegated to distributed small processors.

Where custom, large central processors are employed, more attention will be paid to easing the programming task via design improvements. A "true" stack structure is not unlikely.

The trend toward remote concentrating or semiconcentrating systems with internal intelligence may have profound effects on the controllers of systems upon which they home; their architectures may be somewhat distorted in deference to a potentially heavy remote communication load.

Cache stores, which are small, high-speed memories internal to the processor acting as temporary repositories for a segment of code and data to be repetitively used for a time, are fairly common in high-performance commercial computers. They have not been widely used in telecommunications processors primarily because analyses of "living" call processing code indicated that branches occurred with such extreme frequency that there would be little utility in them. However, program structure and machine architecture greatly influence the dynamic execution behavior of code, and with other structures and architectures, cache stores may become common.

The dropping hardware costs may rekindle interest in triplication and voting, and quadruplication in the form of two matched pairs.

Time division switching systems of the variety being produced are adequate for digitally switching voice and moderate-rate data. For future applications where high-rate data and full bandwidth (and perhaps high-definition) video may need to be switched, however, it is not unlikely that switches will once again become space division, albeit high-speed electronic and digital.

EXERCISES

1. Which system attributes appear to be the most attractive?

2. Suggest an architecture which combines (if feasible) the best selection of attributes of the systems described.

3. Comment upon the likely difficulties in designing a system capable of being successfully marketed internationally.

4. Estimate how long it will take for digital systems to become

    a) significantly numerous;
    b) universal.

REFERENCES

[1] V. K. Rice and J. L. Stella, "No. 2 EAX Type 2A Processor Complex," *Automatic Electric Journal,* January, 1977, p. 226.

[2] P. W. Bassett and R. T. Donnell, "GTD-5 EAX: Recovery and Diagnostic Aspects of a Multiprocessor System," *ISS 81,*.

[3] N. J. Skaperda, "General Digital Switching System," *ISS 76,* p. 223-4.

[4] J. A. Harr et al., "Organization of No. 1 ESS Central Processor," *BSTJ,* Vol. 43, September, 1964, p. 1845.

[5] A. H. Doblmaier and S. M. Neville, "The No. 1 ESS Signal Processor," *"BellLabsRecord,"* April, 1969.

[6] "No. 2 ESS," *BSTJ,* Vol. 48, No. 8, October, 1969.

[7] R. E. Staehler, "1A Processor - A High-Speed Processor for Switching Applications," *ISS 72,* p. 26.

[8] J. S. Buchau and J. Reid, "SP-1 Central Control Complex," *NEC Proceedings,* 1968.

[9] C. Rozmaryn et al., "The E11 Switching System," *ISS 74,* p. 134.

[10] A. E. Pinet, "Introduction of Integrated PCM Switching in the French Telecommunication Network," *ISS 72,* p. 470.

[11] J. Meurling and R. Ericsson, "ARE-A Dual Purpose Switching System for New Installations and for Updating of Exchanges in Service," *ISS 74,* p. 445.

[12] C. G. Larson et al., "Operation Principles and Experience of AXE10 in Integrated Digital Networks," *ISS 81.*

[13] R. A. Lausch et al., "SPESS - An Experimental Stored Program Electronic Switching System," *Automatic Electric Technical Journal,* April, 1968, Vol. II, No. 2, p. 61.

[14] S. H. Liem, et al., "PRX/PDX, the Evolution of a Telephone System," *ISS 76,* p. 411-4.

[15] F. J. Schramel et al., "The Peripheral Control Domain, an All-Digital Intelligent Terminal for Subscribers and Trunks in the PRX-D System," *ISS 81.*

[16] E. Salomonson, "Experience from Multiprocessing in AKE 13," *2nd International Conference on Software Engineering for Telecommunications Switching Systems Proceedings,* February, 1976, p. 92.

[17] K. Sorme and I. Jonsson, "AXE, A Functionally Modular SPC System - System Structure and Operation and Maintenance Features," *ISS 74,* p. 411.

[18] B. Lampe et al., "A Program System for Safeguarding the Availability of the Siemens System EDS," *ISS 74,* p. 141.

[19] F. J. Schramel and A. W. van't Slot, "General Introduction to PRX205," *ISS 72,* p. 349.

[20] R. W. Duthie et al., "C1 EAX - Software and Real-Time Considerations in a Small Stored Program Switching Machine," *ISS 72,* p. 577.

[21] D. Halton, "Hardware of the System 250 for Communication Control," *ISS 72,* p. 530.

[22] T. Tamiya et al., "D-10 Toll and Local Combined Switching System," *ISS 76,* p. 131-3.

[23] A. A. Claxton and M. H. Verbeeck, "Introduction of Next Generation High-Capacity Processors into SPC Systems," *ISS 76,* p. 211-2.

[24] M. Bourbao and J. B. Jacob, "New Developments in E10 Digital Switching System," *ISS 76*, p. 421-4.

[25] J. A. Bloc-Daude et al., "'J2000' Multiprocessor System, Software Architecture Advantages of 'J2000' Distributed Control System," *ISS 76*, p. 422-3.

[26] D. Voegtlen, "CP44 - A Control Complex for Telephone Switching Centers," *ISS 76*, p. 422-4.

[27] H. Baur, "Deutsche Bundespost Prepares Way for New Electronic Switching System," *Siemens Review* XL (1973) No. 10, p. 451.

[28] H. Kunze, "EWS: Germany's Answer to ESS," *IEEE Spectrum*, November, 1975, p. 51.

[29] M. T. Hills and R. P. Loretan, "The Future Direction of SPC Systems," *ISS 76*, p. 412-2.

[30] J. A. Watts, "NX-1E Routing Control Complex," *ICC 70*, p. 27-10.

[31] G. Gattner, "Program Controlled Telephone Switching System Featuring Remote Control and Central Signal Channels," *ICC 70*, p. 27-31.

[32] J. D. Beierle, "10-C Toll Telephone Switching System Control Processor," *ICC 70*, p. 27-19.

[33] P. de Ferra and S. Martinelli, "Basic Features of the Time Division Switching System PROTEO," *ICC 75*, p. 22-20.

[34] R. Galimberti, "PROTEO System: an Overview," *ISS 81*.

[35] J. J. Dankowski et al., "ETS-4 System Overview," *ICC 75*, p. 14-1.

[36] G. F. Dooley and S. Y. Persson, "No. 1 EAX System Objectives and Description," *ICC 73 Record*, p. 10-1.

[37] F. A. Weber, "No. 1 EAX Common Control," *ICC 73 Record*, p. 10-12.

[38] F. Schalkhauser, "ESK 10,000 E - A Family of Systems for Public Telephone Exchanges," *Reports on Telephone Engineering*, 1969, Vol. 213, p. 106.

[39] J. Reines et al., "TCS System Organization and Objectives," *Electrical Communication*, 1973, Vol. 48, No. 4, p. 365.

[40] S. Yazu et al., "D-20 Electronic Switching System," *Japan Telecommunications Review,* Pt. 1, October, 1974, p. 243.

[41] M. Eklund et al., "AXE 10 - System Description," *Ericsson Review,* No. 2, 1976, p. 70.

[42] J. A. DeMiguel, "Telephone Switching Network with Mechanical Latching and Stored Program Control," *Electrical Communication,* Vol. 45, No. 3, 1970, p. 186.

[43] K. Katzeff and U. Jerndal, "The Application of the Stored Program Controlled Switching System AKE in Various Networks," *ICC 69 Record,* p. 35.

[44] L. Noren and S. Sundstrom, "Software System for AKE 13," *Ericsson Review,* No. 2, 1974, p. 35.

[45] M. Ward, "Total Control of Telephone Exchanges by SPC," *General Electric Company Journal of Science and Technology,* Vol. 39, No. 4, 1972, p. 181.

[46] A. J. Cole and K. Piper, "Technology Evolution in System X," *ISS 81.*

[47] D. G. Bryant, "TXE4 - The Evolution of an Analogue System in a Digital Environment," *ISS 81.*

[48] K. Kusunoki et al., "DEX-21 Central Processor," *Rev. of Electrical Communication Laboratory,* Vol. 19, No. 3, March, 1971, p. 235.

[49] S. Malcolm and D. Wagers, "The DMS-10 Evolution Methodology for Operational Stability," *ISS 81.*

[50] J. B. Parker and A. Osterberg, "DMS-100 Family Field Experience," *ISS 81.*

[51] J. C. McDonald, "Five Years' Experience with ITS," *ISS 81.*

[52] T. H. McKinney, "Field Experience with the System CENTURY Digital Central Office," *ISS 81.*

[53] A. E. Joel, Jr., *Electronic Switching: Central Office Systems of the World,* IEEE Press, 1976.

[54] A. E. Joel, Jr., *Digital Central Office Systems of the World,* IEEE Press, 1982.

[55] B. E. Briley and W. N. Toy, "Telecommunications Processors", *IEEE Proceedings*, Vol. 65, No. 9, September, 1977, pp. 1305-1313.

CHAPTER 8

ECONOMICS UNDER REGULATION

8.1 INTRODUCTION

Economics strives to make a science of a field that, while not quite an art, requires prediction of future events in an uncertain world. While the sorcery associated with the term "prediction" may be softened somewhat by substituting the term "estimation," the process is primarily one of (albeit educated) guesswork.

Given assumptions about the future, the process of computing economics parameters is quite mechanical; however, the concepts and viewpoints of economics are at least as valuable to the decision maker as the analytical results.

Regulation has the direct effect of placing constraints upon rate of return, tariffed charges (maximum and minimum), financing sources, etc. Indirect effects include choice of depreciation schedule on capitalized items, encouragement of usage-sensitive charging, distortion of theoretical demand, etc.

8.2 DEMAND ANALYSIS

Demand analysis attempts to determine the amount of a product or service that can be sold over time. It identifies the markets, their sensitivity to price, and the net effect of a new offering upon the firm's sales (e.g., whether it will affect other offerings favorably or unfavorably).

The effect of price upon demand is often plotted as a *demand curve* (see Figure 8-1); note that the price is the ordinate, in keeping with Economics convention. The curve illustrates the reduction of demand with increasing

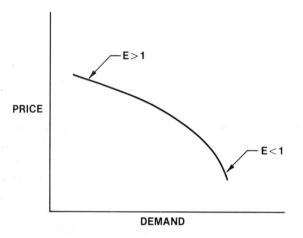

FIG. 8-1. Demand curve.

price, which is characteristic of most marketplace behavior.

Elasticity is defined as

$$E = - \frac{\dfrac{\Delta Q}{Q_0}}{\dfrac{\Delta P}{P_0}},$$

where ΔQ is the change in demand from Q_0 in response to a change ΔP in price from P_0.

The demand for a product is said to be elastic $(E > 1)$ when sales are highly sensitive to price, and inelastic $(E < 1)$ when they are not. A particular brand of television set would tend to experience an elastic demand, while gasoline displays a relatively inelastic demand (in the short term).

A demand curve illustrates expected behavior at a given time; a depiction of demand character over time requires a third dimension, providing a demand-price-time surface.

Another curve of interest is the product/service life-cycle curve, which plots sales (actual or predicted) as a function of time (see Figure 8-2). Such plots typically display an exponential-like initial rise followed by a peak and an exponential-like decay, and tend to be skewed.

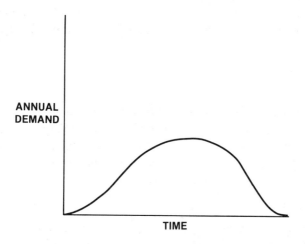

FIG. 8-2. Product life cycle.

Cross-elasticity is defined by an expression identical to that for elasticity except that the price values are for one product or service (the affecting), and the demand values are for another (the affected). The sign of the cross-elasticity will be positive for substitutes (competing products) and negative for complements (mutually benefiting products).

A manufacturer is interested in the product life-cycle curve because he sells products outright (e.g., a switching machine to a Telephone Company [TELCo]). A seller of services that purchased equipment provides (e.g., a TELCo selling telephone service) is interested in the service life-cycle curve.

Competition

Competition has the theoretical effect of producing a discontinuity in the demand curve. Thus, if a competitor offers the same product at a given price, one would expect to capture the entire market by setting a lower price. In practice, there are inevitable differences in the product, or in the way a customer views a product and its manufacturer (sturdier, better looking, better guarantee, better reputation, better known, etc.). Thus, the demand curve tends in practice to display a smooth transition rather than a sharp discontinuity.

Another consideration is the question of legality. The Sherman Act, for example, prohibits unfair competition in the form of subsidization of one product by another. A lower bound is thus effectively set upon allowable prices.

Price Umbrellas

Price structures can be very involved when products and services are complex. If a relatively simple structure is employed, the possibility of a price umbrella exists at points where price and actual costs differ significantly. Such umbrellas may permit competitors to profit in certain market areas.

Tariffs

A rate structure for a service or product provided by a utility is called a *tariff*. Before application, a tariff must by approved by the regulatory commission (or commissions) appropriate to the locale of service or product provision.

Tariffs tend to lag changes in actual costs of service because of the frictional effects of the approval process, so that, for example, the demand estimate for a service impacting an already tariffed class of services may be distorted for some time. Further, since tariffs granted may be different in each of the (approximately 50) jurisdictions, each such jurisdiction may be operating on a different point of a theoretically accurate demand curve.

Demand Estimation

Techniques for estimating demand are numerous, ranging from reasoned guesses through customer surveys to analytical modeling. Survey techniques are simple enough to visualize, while analytical modeling can be quite complex.

For example, a dynamic model may attempt to predict actions that will take place over time as driven by demographic considerations and expected policies.

Another approach to estimating demand is to examine the effects upon rational demand of varying costs of a product to be offered as a replacement for a presently available product. Consider Figure 8-3, which plots the results of computing the cost for each expected application for both product *A* and product *B*, and rank-ordering them according to the magnitude of the difference.

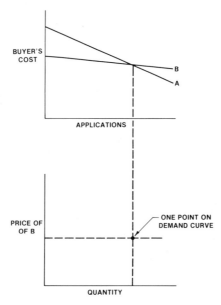

FIG. 8-3. Demand curve construction.

For the corresponding price assumed for product *B*, a rational buyer would prefer *B* over *A* for all cases up to that corresponding to an application for which the costs of utilizing *B* is about that of utilizing *A*. Such a study drives out one point on the demand curve; similar studies varying the price of *B* will provide points to complete the demand curve.

Another curve can be generated as a side benefit to such a series of studies. The area between the curves for *B* and *A* before they intersect represents the "value" of making *B* available at a given price. Therefore a *value curve* can be obtained at the same time as the demand curve. Figure 8-4 depicts a typical value curve; note that it tends to be concave upward, in contrast to demand curve behavior.

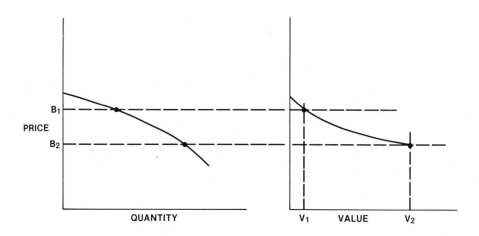

FIG. 8-4. Value curve construction.

8.3 ENGINEERING ECONOMY

Engineering economy provides a means for aiding decisions between alternatives based upon considerations including the time value of money, taxes, depreciation rates, salvage value, operating expenses, availability of capital, etc.

The reference point in time is usually the present, and the Present Worth of Revenue Requirements (PWRR) is the usual basis for comparison (for products with differing service lives, the so-called Levelized PWRR is used). Revenue requirements are recognized by the Federal Communications Commission (FCC) as appropriate considerations in supporting requests to construct facilities. The required revenue is understood to include a reasonable rate of return upon investment.

It is instructive to examine the flow of money in a large company such as that represented by the Bell System. Figure 8-5 depicts the money flow in

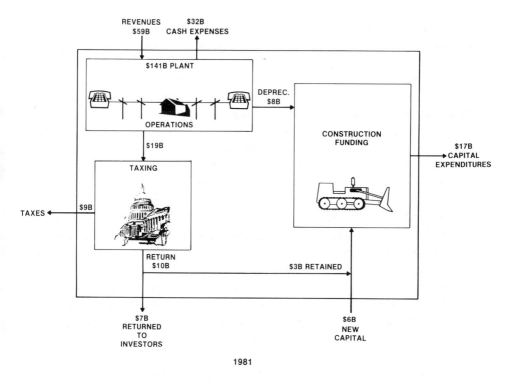

FIG. 8-5. Bell System cash flow: 1981.

1981.

The only influx of cash into the system is from revenues and new capital. Outflow includes cash expenses, income taxes, capital expenditures, and return to investors.

New capital is obtained from sale of stocks (equity capital) and bonds (debt capital). Money returned to investors consists of dividends paid to stockholders, and interest and retirement monies paid to bondholders. A regulated utility is not altogether free to choose its debt-to-equity financing ratio.

Corporation income taxes are limited to about 50 percent of the revenue less expenses, depreciation allowance, and debt interest (certain other considerations, such as investment tax credits, complicate the situation somewhat). In general, some of the earnings are retained, and the remainder is distributed to investors in the form of dividends, bond interest, and bond retirement.

Expenses include wages, property taxes, power costs, etc. Depreciation allowances are intended to permit the original cost of capital equipment to be repaid over a period of time, in theoretical proportion to the reduction in value of the equipment. Thus, ideally, when a piece of equipment reaches the end of its useful life for a firm, the number of dollars accumulated via depreciation allowances over the years, added to the equipment's salvage value, will just equal the original cost. The original investment is therefore available to be returned to the investors, or to be used to purchase a new item of replacement equipment. Of course, due to the time value of money, the present worth of the depreciation allowances will differ substantially from the initial cost of the equipment. Similarly, the price of a replacement for a piece of equipment at the end of life will, in practice, differ from its original price. It is important to recognize that the depreciation allowances are not locked up in inaccessible accounts earmarked only for purchasing replacement capital equipment, but are available for day-to-day use by the business.

The differences between *expensed* and *capitalized* expenditures have subtle as well as obvious facets. For tax purposes, expenses are immediately deductible, while depreciation allowances are deductible in the year they occur. In general, if the treatment of an item is negotiable (with the Internal Revenue Service), there is a tax advantage to expensing it, but at the risk of depressing the apparent present success of the company, reducing the attractiveness of its stock.

In actual accounting practice, individual items are typically lumped into groups, and an average life for the group is used in depreciation calculations. Further, it is usually necessary to maintain two sets of books (quite legally). One set is used for determining the rate-base (first-costs less depreciation reserve) to be employed in computing rates of return for regulation purposes. The other set is used for tax purposes, and employs the more rapid depreciation schedules often allowed by the Internal Revenue Service.

The Time Value of Money

It is convenient to adopt the following widely used notation for expressing the time value of money: given y, to find x for n periods at interest rate i, we denote the operation by

$$(x/y)_n^i,$$

where x and y can take on the roles of

p: present value,

a: annual value,

f: future value.

The convenience of this notation is more fully appreciated when it is recognized that

$$(x/y)_n^i = (x/z)_n^i \times (z/y)_n^i$$

and

$$(x/y)_n^i = 1 \div (y/x)_n^i .$$

More specifically,

$$(p/a)_n^i = \frac{1 - (1 + i)^{-n}}{i} ,$$

$$(a/p)_n^i = \frac{i}{1 - (1 + i)^{-n}} ,$$

$$(f/a)_n^i = \frac{(1 + i)^n - 1}{i} ,$$

$$(a/f)_n^i = \frac{i}{(1 + i)^n - 1} ,$$

$$(p/f)_n^i = \frac{1}{(1 + i)^n} ,$$

$$(f/p)_n^i = (1 + i)^n .$$

Textbooks on economics frequently provide tables of values for the above formulas to speed computation. The advent of hand calculators, and especially of those that have built-in algorithms for the computation of these values, makes tables unnecessary. Further, a survey of such books will show that the tables provided in some are too conservative in their implied estimate of useful interest rates.

It is also convenient to use a pictorial representation of monetary movement (or cash flow) as a function of time. Figure 8-6 provides an example of a computation of the present worth of a machine that would cost $1M now, have a salvage value in ten years of $50K, require $25K per year in expenses, and produce $250K each year in revenues.

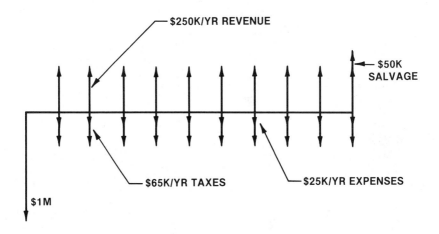

FIG. 8-6. Project cash flow.

Assuming that a straight-line depreciation schedule is applicable, and the tax rate is 50 percent, then, each year, $95K could be deducted for tax purposes, and the yearly tax would be

$$\$(250K - 25K - 95K) \times 50\% = \$65K.$$

The present worth of such an investment, after taxes, would be

$$PW = -1M + 160K\,(p/a)\,_{10}^{10\%} + 50K\,(p/f)\,_{10}^{10\%}$$

$$= -1M + 160K \times 6.14 + 50K \times 0.386$$

$$= -1M + 982.4K + 19.3K$$

$$= \$\,1.7K.$$

As a further illustration of the diagrammatic representation technique, consider the effect of allowing an accelerated depreciation rate, say five years,[1] straight-line (see Figure 8-7). The taxes for the first five years would be

$$\$\,(250K - 25K - 190K) \times 50\% = \$\,17.5K$$

and for the last five years,

$$\$\,(250K - 25K) \times 50\% = \$\,112.5K.$$

The present worth would then be given by

---

1. Relatively unusual but permitted under certain circumstances.

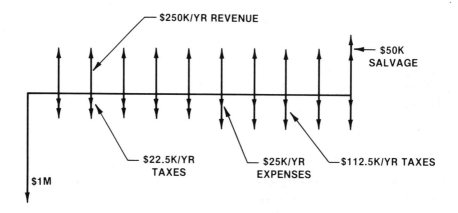

FIG. 8-7. Accelerated depreciation.

$$PW = -1M + 207.5K\,(p/a)_5^{10\%} + 112.5K\,(p/a)_5^{10\%}(p/f)_5^{10\%} + 50K\,(p/f)_{10}^{10\%}$$

$$= -1M + 207.5K \times 3.79 + 112.5K \times 3.79 \times 0.621 + 50K \times 0.386$$

$$= -1M + 786.4K + 264.8K + 19.3K$$

$$= \$70.5K.$$

Aside from further illustrating application of the cash-flow diagram, the effect of the choice (when there is one) of depreciation schedule becomes clear: more rapid depreciation is advantageous. Other depreciation methods include double declining balance (twice the straight-line rate), sum of year's digits (remaining years divided by the sum of the years of life), and sinking fund (little used because depreciation is depressed in early years). Immediate write-off (equivalent to expensing) would, of course, be economically ideal.

Income taxes include those of local government as well as Federal levies, and one component is often deductible before computing the other. A composite income tax figure representing the net effect of the several taxes is therefore often used to simplify computation in economic studies; it is typically very close to 50 percent.

If a minimum acceptable rate of return has been established for a firm relative to new project ventures, then the present equivalent of project cash flows may be employed in comparing candidate projects at that interest rate. Annual equivalent values are also employed (future equivalents would yield the same relative results, but are less frequently used).

The minimum rate of return may be expressed as a before- or after-taxes figure, the latter being generally more meaningful because debt capital interest and depreciation schedule choice influence taxes substantially, and therefore influence the relation between before- and after-tax figures.

If the rate of return required has not been established, then a *prospective rate of return* can be computed. It is the interest rate at which the present equivalent of cash flows is zero. That is, it is the interest rate which, if the same investments were made (say, in bonds), would yield the identical yearly equivalent return.

In the first example given, the present equivalent was negligibly small compared to the investment, so that the prospective rate of return would in fact be approximately 10 percent. In the second example, since the present equivalent was nonzero and positive, the prospective rate of return must have been greater than 10 percent.

Clearly, if a choice must be made between two projects, the one with the highest prospective rate of return is preferred. However, if neither project is necessary, and neither has a rate of return greater than or equal to that established as a threshold, both should be rejected. If both projects have a prospective rate of return greater than the threshold, then they should both be accepted so long as adequate funds are available for investment. Such considerations are readily extendible to arbitrary numbers of competing projects.

A word of caution is in order relative to the calculation of the prospective rate of return: the cash-flow equation may, in complex situations, have more than a single root, corresponding to a multiplicity of return rates; care must be exercised to choose the root that is appropriate.

EXERCISES

1. The prize in a lottery is $1M to be given in equal payments over 20 years. How much money would have to be placed in a bank now to provide such an annuity? (Cost of money = 10%.)

2. A project will cost $1M now and $100K per year for its 20-year life. It will produce income of $75K per year for the first 10 years and $200K per year for the last 10 years. For a 10% interest rate and a $250K salvage value, is the project worth undertaking? Neglect taxes.

3. A project has a first cost of $100K and $10K salvage value at retirement in 10 years. Income is expected to be $30K per year and expenses $10K. Income taxes are 50%, and straight-line depreciation will be used. Find the present worth of cash flow (cost of money = 10%).

4. What would be the effect of the inflation rate becoming equal to the cost of money?

READING LIST

G. W. Smith, *Engineering Economy: Analysis of Capital Expenditures*, Iowa State University Press, 1973, 2nd Edition. (Unusually good.)

AT&T, *Engineering Economy, A Manager's Guide to Economic Decision Making*, Third Edition, McGraw-Hill, 1977 (Very pertinent.)

T. J. Morgan, *Telecommunications Economics*, Technicopy, Ltd, 1975.

J. R. Cavada, *Intermediate Economic Analysis for Management and Engineering*, Prentice-Hall, 1971.

H. Bierman and S. Smidt, *The Capital Budgeting Decision*, Macmillan, 1971.

GLOSSARY

GLOSSARY

ACD: Automatic Call Distributor - A system that automatically distributes calls to operator pools (providing services such as intercept and directory assistance), to airline ticket agents, etc.

Administration: The tasks of record-keeping, monitoring, rearranging, predicting need for growth, etc.

AIS: Automatic Intercept System - A system employing an audio-response unit under control of a processor to automatically provide pertinent information to callers routed to intercept.

Alert: To indicate the existence of an incoming call, e.g., by ringing.

AMA: Automatic Message Accounting - The capability within an office for automatically recording call information for billing purposes.

AMPS: Advanced Mobile Phone System - A Bell System-designed mobile telephone system allowing the reuse of frequencies over relatively short distances.

ANI: Automatic Number Identification - Often pronounced "Annie," a facility for automatically identifying the number of the calling party for charging purposes.

Appearance: A connection upon a network terminal, as in "the line has two network appearances."

Assembler: A program which converts (assembles) assembly-level language statements (source code) into machine-level binary (object code). It need not be implemented on the machine for which the program to be converted is written. Assembly is, in general, a one (statement)-to-one conversion process.

Attend: The operation of monitoring a line or an incoming trunk for off-hook or seizure, respectively.

Audible: The subdued "image" of ringing transmitted to the calling party during ringing; not derived from the actual ringing signal in later systems.

Backbone Route: The route made up of final-group trunks between end offices in different regional center areas.

BHC: Busy Hour Calls - The number of calls placed in the busy hour. *See also* Busy Hour.

Blocking: The ratio of unsuccessful to total attempts to use a facility; expressed as a probability when computed a priori.

Blocking Network: A network that, under certain conditions, may be unable to form a transmission path from one end of the network to the other. In general, all networks used within the Bell System are of the blocking type.

Blue Box: Equipment used fraudulently to synthesize signals, gaining access to the toll network for the placement of calls without charge.

BORSCHT Circuit: A name for the line circuit in the central office. It functions as a mnemonic for the functions that must be performed by the circuit: Battery, Overvoltage, Ringing, Supervision, Coding (in a digital office), Hybrid, and Testing.

Bus: A logic data path with a multiplicity of potential sources or a multiplicity of potential destinations for information upon it (or both); usually several bits wide though often diagrammatically shown as a single line with its width beside it in parentheses.

Busy Hour: A one-hour period selected for a given office to typify the annually recurring hour during which the offered traffic load is a maximum.

Busy Signal: (Called-line-busy) An audible signal which, in the Bell System, comprises 480 Hz and 620 Hz interrupted at 60 IPM.

Bylink: A special high-speed means used in crossbar equipment for routing calls incoming from a step-by-step office. Trunks from such offices are often referred to as "bylink" trunks even when incoming to noncrossbar offices; they are more properly referred to as "dc incoming trunks." Such high-speed means are necessary to assure that the first incoming pulse is not lost.

Cable Vault: The point in a central office where external plant cables enter the building.

Call Store: The memory used in an ESS to contain transient data on calls in progress or being processed, Recent Change information, and the network map.

CAMA: Centralized Automatic Message Accounting - Always pronounced: rhymes with Alabama. A centralized form of AMA, usually in a larger office serving a number of end offices that may be too small to justify AMA features. Since it is necessary to supply information about the calling line to the CAMA office, an office that is homing upon a CAMA office must have ANI capability unless the CAMA office is of the "operator identified" variety, in which the calling line number may be manually obtained.

CCIS: Common Channel Interoffice Signaling - Signaling information for trunk connections over a separate, nonspeech data link rather than over the trunks themselves. First being applied between toll offices, eventually throughout the system.

CCITT: International Telegraph and Telephone Consultative Committee - An international committee that formulates plans and sets standards for intercountry communication means.

CCS: (100 Call Seconds) - A measure of traffic load obtained by multiplying the number of calls per hour by the average holding time per call expressed in seconds, and dividing by 100. Often used in practice to mean hundred call seconds per hour with "per hour" implied; as such, it is a measure of traffic intensity. *See also* Traffic Intensity.

CDO: Community Dial Office - A small, usually rural office typically (essentially universally at this time) served by step-by-step equipment. Characterized by low calling rates and chiefly intraoffice traffic.

Central Office: The central office comprises a switching network and its control and support equipment. Occasionally improperly used to mean "office code." The term "office" should not be confused with the usual layman's interpretation.

Centrex: A service comparable in features to PBX service but implemented with some (Centrex CU) or all (Centrex CO) of the control in the central office. In the latter case, each station's loop connects to the central office.

Churning: The phenomenon, especially prevalent in large metropolitan areas, of frequent customer movement, requiring many telephone installations per net-gained customer. Ratios as great as 10:1 are seen in Manhattan.

Clos Network: A class of multistage nonblocking network capable of implementation with fewer crosspoints (above trivial sizes) than a single-stage network. A man's name, *not* an acronym.

CO: See Central Office.

Community of Interest: A group of subscribers more or less in the same geographical area who share sufficient common interests to be more likely to make calls within the group than out of it.

Compiler: A program that converts (compiles) another program, written in a high-level language, into a machine-understandable form (usually binary). In general, a different compiler program must be written (implemented) for each machine upon which programs written in the high-level language are to be run. Compilation is, in general, a one (statement)-to-many conversation process.

Concentration: Providing to the inlets to a network, access to a lesser number of links to another switching stage.

Crossbar: See X-Bar.

Crosspoint: The means for interconnecting two channels on a space-division basis; the channel may consist of one, two, three, or four wires. Examples include the ferreed switch, for which two or four individual dry reed switches are regarded as a crosspoint.

Customer Loop: The wire pair connecting a customer's station to the central office.

Cutover: The act of disconnecting an existing switching machine and connecting in its place another, usually newer machine. Typically performed in the wee hours when telephone traffic is near nil.

DDD: Direct Distance Dialing - Dialing without operator assistance over the nationwide intertoll network.

Debt Capital: Funds obtained by a company through selling bonds.

Decorrelator: Means for distributing the TDM traffic incoming to a No. 4 ESS office to prevent correlated traffic on one trunk group from causing abnormal blocking.

Direct Trunk Group: A trunk group that is a direct connection between a given originating and a given terminating office.

Distribution: The process of distributing transmission paths to various portions of the next stage of switching in a network without, in general, entailing either concentration or expansion.

Drop: The wire-pair extending from the distribution cable to the subscriber premises.

DTL: (Diode Transistor Logic) - A logic type which performs diode logic (AND or OR) followed by a one-transistor inverter yielding a net NAND or NOR. Used in No. 1 ESS; nominal delay $\simeq$ 35 ns.

Emulation: The process of simulating the actions of one machine upon another by programming its microprogram memory such that the emulating machine accepts and executes programs written for the emulated machine. The classical definition permits some software activity (at machine-language level) as well as firmware (the term for microprograms) in performing this function.

End-Marked Network: A network so constituted that a path is cut through as a result of merely "marking" an inlet and an outlet terminal (usually by applying a potential). No present Bell System network is of this variety, but the Morris, Illinois, pre-No. 1 ESS field trial used such a network.

End Office: A class 5 central office typically serving only customer lines.

Engineer: (v.t.) To specify the number or arrangement of facilities (trunks, networks, etc.) to meet a given grade of service criterion (e.g., to engineer a trunk group for .01 blocking probability).

Entity: See Central Office.

EOTT: End Office Toll Trunking - Trunking between end offices in different toll center areas.

Equity Capital: Funds obtained by a company through selling stock.

Equivalent Random Method: A method due to Wilkinson that allows computation of the number of servers needed to accommodate rough traffic.

Erlang: This is a measure of telephone traffic intensity equivalent to the average number of simultaneous calls. Alternatively, it is the total circuit usage in an interval of time divided by that interval. Thus 1 erlang equals 3600 call seconds per hour or 36 CCS per hour. The term erlang is used in honor of A. K. Erlang; it is *not* an acronym.

ESB: Emergency Service Bureau - A centralized agency to which 911 "universal" emergency calls are routed.

ESS: Electronic Switching System - A generic term used to identify as a class, stored-program switching systems such as the Bell System's No. 1, No. 2, No. 3, No. 4, or No. 5.

ETS: Electronic Translation System - An electronic replacement for the card translator in 4A Crossbar systems. Makes use of the SPC 1A Processor.

Expansion: Providing to the links entering a switching stage, access to a larger number of outlets.

False Start: An aborted dialing attempt.

Fast Busy: (often called reorder) - An audible busy signal interrupted at twice the rate of the normal busy signal; sent to the originating station to indicate that the call blocked due to busy equipment.

Ferreed: The network switch element used by No. 1 and No. 2 ESS; two or four reed switches in individual glass capsules are enclosed by magnetically hard magnetic material (remendur, not ferrite) and windings capable of magnetizing the structure so that the switches close, or in effect, demagnetizing it so that they open; a pulse of about 9 amperes is applied for 300 microseconds (the switches then take several milliseconds to respond).

Ferrite Sheet: An electronically alterable memory consisting of an apertured sheet of square-loop ferrite material with printed conductors and plated-through holes; used for the "call stores" of No. 1 and No. 2 ESS.

Ferrod: The scanning sense element used by No. 1, No. 2, and No. 3 ESS; a rod of magnetically soft ferrite material is wound with the line conductor and a drive and sense wire; acting as a saturable transformer, e.g., when a line goes off-hook, the current drawn saturates the material and prevents an interrogating (scanning) pulse from being sensed by the pickup winding. Ferrods of differing sensitivity are used for monitoring lines, trunks, and junctors.

Final Trunk Group: The trunk group to which calls are routed when available high-usage trunks overflow; these groups generally "home" on an office next highest in the hierarchy.

Firmware: A term applied to control programs written for the microprogram control store of a microprogrammed machine; term intended to imply a middle ground between hardware and software.

Folded Network: A network which has lines and trunks on the same side, requiring that the network, in effect, be "folded" about its center such that a call passes through it twice; the concentration stage effectively functions as an expansion stage in the opposite direction.

Foreign Potential: An improper voltage impressed upon a line or trunk, e.g., due to a power line falling across a cable.

Frogging: The practice of positionally interchanging the lines of an open-wire pair to reduce induced noise.

Full-Access Network: A network that permits any inlet to access any outlet.

Full Group: A trunk group that does not permit rerouting off-contingent foreign traffic; there are seven such offices.

Glare: The situation that occurs when a two-way trunk is seized more or less simultaneously at both ends.

Hamming Code: A checking and correcting code that is in the class of cyclical redundancy codes.

Heat Coil: A protective device used to open a line or trunk circuit when excessive current is drawn due to some malfunction.

High-Usage Trunk Group: The appellation for a trunk group that has alternate routes via other similar groups, and ultimately via a final trunk group to a higher ranking office.

HILO: A technique of driving a network input port with a high-impedance source, and terminating an output port with a low-impedance sink. Useful for permitting low-noise, path-impedance-insensitive, unbalanced transmission through a network.

Holding Time: The duration of a call expressed in any time unit; also denotes the duration of holding of any resource, though it may be in use for only a fraction of the total call duration.

Home (On): To be one of the offices connected via a final trunk group to an office higher in the hierarchy (e.g., "end office A homes on class 4 office B").

Hopper: A buffer area in ESS call store (many such used).

Hybrid: The Hybrid Transformer and matching impedance used in the telephone station to convert from two-wire to four-wire transmission. Such circuits are also used elsewhere in the network for the same purpose.

IGFET: Insulated Gate Field Effect Transistor - A unipolar device with very high input impedance and low power dissipation.

Intensity (Traffic): Offered traffic load per unit time.

Intercept: The agency (usually an operator) to which calls are routed when made to a line recently removed from service, or in some other category requiring explanation. Automated versions (AIS) with automatic voice response units are growing in use.

Interoffice: Between two offices (e.g., an interoffice call).

Interrupt: A facility in ESS machines (and many computers) that can cause a program in progress to cease in deference to a program more urgently required. After the second program has completed (assuming it has not been interrupted by a need still higher in the hierarchy) the first is resumed.

Intraoffice: Within the same office (e.g., intraoffice traffic).

ITU: International Telecommunications Union.

Junctor: A wire or circuit connection between networks in the same office. The functional equivalent to an intraoffice trunk.

LAMA: Localized Automatic Message Accounting - Always pronounced: rhymes with Alabama. The capability for AMA in a local office.

LAMA-C: Computerized AMA for No. 5 Crossbar systems.

Limited-Access Network: A network that restricts access to some outlets.

Line-Feed Inductor: A split-winding primary transformer used in the line circuit to eliminate longitudinal signals and to supply battery without interference among lines.

Link: One of the transmission paths used to make connections between successive stages of switching in a network.

Load Balancing: The process of rearranging customer connections to the first-stage switch of the network to distribute traffic more equitably.

Longitudinal: An unwanted signal impressed equally upon both wires of a wire-pair transmission line. It can be eliminated by careful balancing.

Lost Calls Cleared: A requester behavior in a traffic system whereby the requester leaves the system if all servers are found busy.

Lost Calls Delayed: A requester behavior in a traffic system whereby the requester remains in the system until served regardless of blockage.

Lost Calls Held: A requester behavior in a traffic system whereby the requester remains in the system whether blocked or not (including the case of a portion of the time spent waiting to be served), for a total of an average holding time.

Loud Ringer: A special ringer supplied upon request for locations with extreme background noise.

Main Distributing Frame: The frame that supports the jumpers connecting lines and trunks to the office network.

Main Station: A term that differentiates between the individual stations that may exist and the actual number of lines in existence. Thus, for example, if a customer has several stations that are used for the purpose of extensions, though many may exist, they will count collectively as a single main station.

Mark: (v.t.) To apply an appropriate potential to a shared entity to indicate to competing users that it is not idle, e.g., to mark the trunk busy. Also, in electronic systems, to change a memory element's state to indicate busy/idle status of some facility.

Marker: A controller (one of several) in a crossbar office.

MDF: See Main Distributing Frame.

MF: See Multifrequency.

Morris (Illinois): Site of the field trial predecessor to No. 1 ESS (also known as the "Morris System").

Multifrequency (MF): The method of signaling over a trunk making use of the simultaneous application of two out of six possible frequencies.

Multiple: (n) The parallel connection between like appearances of an entity (e.g., the multiple between the appearances of customer's jack on several switchboards); (vt) the act of making such connections.

Multiprocessor: A collection of processors interconnected in such a way they cooperate (in some sense) in performing the required task.

Network: The collection of switching elements and interconnections which, when proper elements are activated, is capable of supporting a multiplicity of distinct transmission paths for voice or signal transmission.

Nonblocking Network: A strictly nonblocking network is always capable of forming an independent transmission path from any inlet to any outlet. A network that is nonblocking in the wide sense will not block so long as paths are set up in accordance with an appropriate set of rules.

NPA: Numbering Plan Area.

Object Code: The code produced by an assembler or compiler in converting (assembling or compiling) the input source code to a machine- understandable form. In binary when run, though often expressed in other forms (octal, hexadecimal) for easy reading.

Occupancy: Percentage of time or probability of being busy.

One-Way Trunk: A trunk that can be seized at only one end.

ONI: Operator Number Identification - The use of an operator in a CAMA office to verbally obtain the calling number of a call originating in an office not equipped with ANI.

OR: See Originating Register.

Order Wire: An archaic system for communication between operators at different offices. Often referred to as the ancestor of CCIS.

Originating Register: A device (one of several) associated with a crossbar machine, which is connected to an originating line after off-hook has been detected, supplies dial tone and remains with the line until dialing is completed; also a group of locations in ESS memory used for retaining similar information as it is collected piecemeal by the control from a dial pulse or TOUCH-TONE receiver.

Origination: The act of a subscriber going off-hook in anticipation of dialing.

PABX: Private Automatic Branch Exchange - An automatic telephone office serving a private customer, e.g., a business, school, etc.; typically, access to the outside telephone network is provided. The PBX services provided often include features not commonly available to the general telephone customer, e.g., conference calling, call waiting, etc.

Packing: The compacting of the carbon granules in a carbon transmitter. Usually a result of age or disuse. Results in lessened sensitivity. Also an algorithm for path hunt in a network: search for paths in the most congested part of the network first.

PAM: Pulse Amplitude Modulation - A modulation technique that represents a signal with a succession of pulses whose amplitude is in proportion to that of the original signal at the corresponding sampling times. Used, e.g., in No. 101 ESS's time-division network.

Panel: An electromechanical office type using a gross-motion driven switch. Installed during the '20s and '30s exclusively in "downtown" metropolitan offices. Rapidly replaced by No. 1 ESS; extinction occurred in 1982.

Panel Call Indicator: A method originating with panel equipment for signaling to operator boards.

PBX: See PABX.

PCI: See Panel Call Indicator.

PCM: Pulse Code Modulation - A modulation technique that represents a signal with a succession of groups of pulses, each group comprising a binary representation of the original signal's amplitude at the corresponding sampling time. Widely used in transmission, employed, e.g., in association with the time-division network for No. 4 ESS.

Permanent Signal: A sustained off-hook condition without activity (no dialing or ringing or completed connection); such a condition tends to tie up equipment, especially in earlier systems. Usually accidental, but sometimes used intentionally by customers in high-crime- rate areas to thwart prospective burglar calls.

Piggy-Back Twistor: See Twistor.

POL: Problem Oriented Language - A high-level language such as FORTRAN.

POTS: Plain Old Telephone Service; basic service without the "frills" such as custom calling - Usually pronounced.

PPM: Pulse Position Modulation - A modulation technique that represents a signal with a succession of pulses whose position in time relative to a standard position is in proportion to the original signal's amplitude at the corresponding sampling time.

Price Umbrella: The situation that occurs when a tariff is written such that some services have charges that are significantly greater than compensatory, leaving open the possibility of a competitor offering the services for less while still making a profit.

Pushdown Stack: Conceptually, a group of registers which can pass information back and forth between adjacent members; usually a LIFO (last-in, first-out) configuration (e.g., the stack used in Hewlett-Packard's current line of hand-held calculators). Most commonly implemented as a group of adjacent memory locations and an address counter which keeps track of the location of the last entry. Used, e.g., in No. 2 and No. 3 for storing subroutine return addresses.

PWM: Pulse Width Modulation - A modulation technique that represents a signal with a succession of pulses whose width relative to a standard width is in proportion to the original signal's amplitude at the corresponding sampling time.

Rearrangeable Network: A network in which it is permissible to move a subset of the paths that are up to accommodate a new path.

Receiver: A shared means (one of many) for detecting incoming information one character at a time; includes dial-pulse receivers and TOUCH-TONE receivers (the latter can often receive dial pulses as well) for originating lines, and MF receivers for incoming trunks.

Register: A group of storage elements capable of storing a number; in a stored-program processor it would consist of flip-flops, in a crossbar marker, of latching relays; also used to denote an equivalent telephone function performed in the memory of an ESS processor. *See also* Originating Register.

Register Progressive (System): A system that records dialed digits in a common receptacle (register), then spills them forward into a progressive network (e.g., step-by-step). Advantages include the ability of retrial and compatibility with TOUCH-TONE signaling.

Remreed: A newer network switch element similar to the ferreed but having reed blades composed of remanent magnetic material, rendering the structure much more sensitive and requiring less external magnetic material, fewer windings, and less current for activation.

Reorder: See Fast Busy.

Resistance Design: Choice of loop pair gauge and loading on the basis of loop resistance as a criterion indicative of quality of transmission, signaling, and supervision. Depending upon the station set, typically in the range of 1000 to 1500 ohms.

Revertive Pulsing: A means employed by panel offices for interoffice communication characterized by a feedback mode of operation whereby a destination office sends pulses back to the sending office until a stop signal from the latter indicates that the number of pulses sent back corresponds to the number intended to be transmitted forward. Chosen because of the nature of the equipment. Used also by No. 1 Crossbar for both interoffice and intraoffice communication; chosen because of the larger number of panel offices in existence at the time.

ROTL: Remote Office Test Line - A means for remotely testing trunks.

Rough Traffic: Traffic distribution with a variance greater than the mean.

RSS: Remote Switching System - A small dependent line concentrating system with microprocessor control and intraswitching capability using PNPN switches.

RTA: Remote Trunk Arrangement - An extension to the TSPS system permitting its services to be provided up to 200 miles from the TSPS site.

RTL: Resistor Transistor Logic - A logic type that makes use of series resistors at the inputs of a gate and a shunt resistor to a suitable potential such that the threshold potential for turning on the transistor at its base is reached when one or more input is at a logical "1;" the inversion provided by the common emitter configuration yields a net NOT OR or NOR logic function. Used in the No. 2 ESS system; nominal delay $\simeq$ 35 ns.

Seize: (v.t.) To make connection to and so signify (or mark) by applying an appropriate potential, e.g., to seize the first idle trunk. By analogy in ESS, to make use of and mark busy in memory a block of memory such as an Originating Register.

Sender: Nominally a means for buffering, then spilling in the proper signaling "language," information onto an outgoing trunk; also used to denote means for receiving and buffering with associated control, e.g., a step-by-step office that has a sender (or has been "senderized") is capable of receiving TOUCH-TONE signals from a line and converting them to dc pulses for driving the step-by-step system. No. 1 Crossbar offices use two types, originating and terminating; the first receives dialed digits and transmits them to the second or to a distant office; the second receives them from the first or from a distant office.

SF: Single Frequency. A signaling method used for trunks: 2600 Hz is impressed upon idle trunks.

Sidetone: The replica of the signal generated at the microphone presented at the receiver of the same station. Though it can be eliminated via the hybrid circuit, a small measure of sidetone is desirable.

Signal Processor: An auxiliary and distinct stored-program processor used in association with No. 1 ESS in large offices to relieve it of the task of dealing with the periphery. Also, wired-logic processors performing a similar function as a part of the No. 4 ESS.

Single Frequency: See SF.

Smooth Traffic: Traffic distribution whose variance-to-mean ratio is less than or equal to one.

Software: The programs written for a stored-program machine; the term encompasses programs written to be run on an ESS machine as well as supporting programs written to be run on general-purpose computers.

Source Code: A program in its "as written" form, e.g., the collection of FORTRAN statements or assembly language lines generated by a programmer.

SP: See Signal Processor.

SPC 1A: The stored-program processor employed in the TSPS and 4A/ETS systems.

Station: The telephone instrument used as the means of communication by the customer to the system and to other customers.

Step-by-Step: See S×S.

Strowger Switch: (Strojer Switch) - *See* S×S.

STS Switch: A digital network that has a Space-Time-Space switching architecture.

Subroutine: A program segment of sufficiently common value to justify its use by a multiplicity of "client" programs that branch (or jump or transfer) to it, and are branched back to after its execution is completed.

Subset: Telephone instrument or station.

Supervise: To monitor the status of a call.

Switch Train: The network switching apparatus.

S×S: (Step-by-Step or Strowger switch) - An electromechanical office type utilizing a gross-motion stepping switch as a combination network and distributed control. The first, and most widely used, automatic central office type: originated in the 1890s and still much in evidence.

Talking Battery: An electrically quiet battery isolated from other battery uses to allow good voice communication quality.

Talkoff: The phenomenon of accidental synthesis of a machine-intelligible signal by human voice causing an unintended response.

Tandem (Office): Principally, "local tandem" offices that switch interlocal, 2-wire traffic and have no standing in the hierarchy. Applied as well to certain toll systems. The term *Transit* is frequently used as a synonym in non-US systems.

Tariff: The published rates, charges, rules, and regulations governing the provision of telecommunications services.

Time-Slot Interchanger: Means for interfacing one or more incoming time-division channels to a plurality of outgoing time-division channels, permitting movement from the *i*th time slot of an incoming channel to the *j*th time slot of an outgoing channel.

Tip-Ring-Sleeve: The tip, ring, and sleeve leads are named from their early implementation in manual offices. The tip and ring leads correspond to the two wires which connect the customer with the central office. The so-called sleeve lead exists only within the network of the central office itself and does not exist in all networks (particularly the more modern ones).

Toll: (adj.) - Having to do with calls for destinations outside the local service area of the calling station. Calls for which detail billing is provided.

Toll Center: A class 4 office offering operator assistance for incoming calls.

Toll Point: A class 4 office not offering operator assistance for incoming calls.

Traffic: The aggregate of calling activity in a telecommunications system.

Traffic Density: The number of simultaneous calls at a given moment.

Traffic Intensity: The product of average holding time and the calling rate; expressed in erlangs or CCS per hour.

Traffic Mix: The mixture of intraoffice, incoming, and outgoing traffic experienced by an office.

Transit: See Tandem.

Translation: The process of drawing correspondences via data retrieval between directory numbers and equipment addresses, and services the corresponding customer is paying for (e.g., TOUCH-TONE for a calling line; call waiting for a called line).

Trunk: A path between central offices; in general 2-wire for interlocal, 4-wire for intertoll.

Trunk Splintering: The effect of increasing the number of trunk groups (and reducing their size) from a wire center as a result of using two or more switching machines to replace an existing one that has exhausted. The efficiency of the splintered groups is thereby reduced, requiring more total trunks.

TSI: See Time Slot Interchanger.

TSPS: Traffic Service Position System - A system that provides, under stored-program control, efficient operator assistance for toll calls. It does not switch the customer, but provides a bridging connection to the operator.

TST Switch: A digital network that has a Time-Space-Time architecture.

Twistor: A form of read-only magnetic memory employed in the "program" stores of No. 1 and No. 2 ESS; the twistor elements, composed of a spiral wrap of magnetic ribbon about a conducting wire, are used to sense the remnant state of small magnets which are encoded off-line. The piggyback twistor, which is electronically writable, has an additional wrap of magnetic material that is magnetically "hard" and is used with the SPC 1A Processor.

Two-Way Trunks: Trunks capable of being seized at either end.

Wire Center: Originally the centroid of the lines to be served in a given geographical area. Often loosely used to mean large, multioffice buildings which house one or more switching systems serving one or more office codes.

X-bar: (Crossbar) - An electromechanical office type utilizing a "fine-motion" coordinate switch and a multiplicity of central controls (called markers). There are five varieties:

No. 1 Crossbar: Used in large urban office application; introduced in 1938.

No. 3 Crossbar: A small system introduced in 1974.

No. 4A/4M Crossbar: A 4-wire toll machine; introduced in 1943.

No. 5 Crossbar: A machine originally intended for relatively small suburban applications; capable of large office use; introduced in 1948.

Crossbar Tandem: A machine used for interlocal office switching.

911 Service: A universal emergency number service in existence since 1968 and being expanded in scope.

1A Processor: The stored-program successor to the No. 1 ESS processor, upward compatible therefrom. Used in Nos. 1 and 4 ESS.

INDEX

INDEX

A

ACD: Automatic Call Distributor, 82, 229
Administration, 229
AIS: Automatic Intercept System, 84, 229
AKE (11, 12, 13), 203
Alert, 20, 229
AMA: Automatic Message Accounting, 67, 95, 229
AMPS: Advanced Mobile Phone System, 115, 229
ANI: Automatic Number Identification, 67, 114, 229
Appearance, 18, 229
ARE, 206
Assembler, 187, 230
Attend, 3, 230
Audible, 10, 21, 230
AXE, 206

B

Backbone Route, 94, 230
BHC: Busy Hour Calls, 147, 230
Blocking, 123, 230
Blocking Network, 127, 230
Blue Box, 108, 230
BORSCHT Circuit, 24, 230
Bus, 230
Busy Hour, 139, 141, 145, 230
Busy Signal, 10, 231
Bylink, 37, 231

C

Cable Vault, 7, 231

Cache Memory, 194
Call Store, 60, 231
CAMA: Centralized Automatic Message Accounting, 94, 97, 231
CCIS: Common Channel Interoffice Signaling, 106, 231
CCITT, 189, 231
CCS, 140, 231
CDO: (Community Dial Office), 184, 232
Central Office, 18, 232
Centrex, 232
Charging, 95
CHILL, 189
Churning, 232
Clos Network, 120, 124
CO: Central Office, 18, 232
Code Conversion, 69
Community of Interest, 148, 232
Compiler, 188, 232
Concentration, 121, 232
Control Switching Point, 91
Crossbar: X-Bar, 41, 247
    No. 1 Crossbar, 42
    No. 3 Crossbar, 42
    No. 4 Crossbar, 69
    No. 4A Crossbar, 69
    No. 5 Crossbar, 47
    No. 5A Crossbar, 49
    Crossbar Tandem, 42
Crosspoint, 122, 233
Customer Loop, 233
Cut over, 233

D

DDD: Direct Distance Dialing, 69, 233
Debt Capital, 221, 233
Decoder, 38
Decorrelator, 74, 233
DEX, 207

Dial Tone, 10
Direct Progressive, 33
Direct Trunk Group, 233
Distribution, 121, 233
DMS (10, 100, 200), 204
DNHR, 95
Drop, 3, 233
DTL: (Diode Transistor Logic), 54, 233

E

EAX (1, 2, 3, 5), 202
Emulation, 68, 191, 233
End Marked Network, 23, 51, 234
End Office, 31, 91, 234
Engineer, 234
Entity, 18, 234
Equivalent Random Method, 166, 234
Equity Capital, 221, 234
Erlang, 140
ESB: Emergency Service Bureau, 114
ESK, 205
ESS: (Electronic Switching System), 49
   No. 1 ESS, 50
   No. 1A ESS, 63
   No. 2 ESS, 65
   No. 2A ESS, 66
   No. 2B ESS, 68
   No. 3 ESS, 67
   No. 4 ESS, 71
   No. 5 ESS, 68, 76
ETS, 70, 234
EWSA, D, 205
Expansion, 121, 235
Extreme Value Engineering, 168
E10, (11, 12), 205

F

False Start, 148, 235
Fast Busy: (often called reorder), 235
Ferreed, 51, 235
Ferrite Sheet, 60, 235
Ferrod, 54, 235
Final Trunk Group, 91, 235
Firmware, 235

Folded Network, 54, 236
Foreign Potential, 4, 236
Frogging, 4, 236
Full Access Network, 121, 236
Full Group, 165, 236

G

Glare, 236
Gross Motion, 33

H

Hamming Code, 59, 236
H-Diagram, 147
Heat Coil, 4, 236
Hierarchy (Toll), 91
High-Usage Trunk Group, 91, 236
Holding Time, 140, 236
Home (On), 64, 91, 237
Hopper, 237
Hybrid, 6, 123, 237

I

IGFET: Insulated Gate Field Effect
  Transistor, 68, 237
Infix, 193
Intensity (Traffic), 140, 237
Intercept, 84, 237
Interoffice, 237
Interrupt, 194, 237
Intraoffice, 237
ITS, 203
ITT, 207
ITU: International Telecommunications Union,
  113, 235, 237

J

Junctor, 13, 237

L

LAMA, 237
LAMA-C, 237
Limited Access Network, 165, 238
Line-Feed Inductor, 8, 238
Line Load Control, 147
Link, 121, 238
Load Balancing, 123, 238
Longitudinal, 238
Lost Calls Cleared, 150, 154, 238
Lost Calls Delayed, 150, 157, 238
Lost Calls Held, 150, 153, 238
Loud Ringer, 238

M

Main Distributing Frame, 26, 238
Main Station, 238
Mark, 238
Marker, 42, 48, 239
MDF, 26, 239
Metaconta, 207
MF, 104, 239
Microprogram, 68, 184
Morris (Illinois), 50, 239
Multifrequency (MF), 104, 239
Multiple, 239
Multiprocessor, 192, 239

N

Network, 13, 119, 239
Nonblocking Network, 15, 123, 239
NPA, 70, 109, 239
NX-1E, 203

O

Object Code, 239
Occupancy, 240
ODD, 111
One-Level Limit Rule, 93
One-Way Trunk, 240
ONI, 240
OR, 195, 240
Order Wire, 4, 107, 240
Originating Register, 48, 195, 240
Origination, 3, 240

P

PABX, 240
Packing, 169, 240
PAM, 137, 240
Panel, 37, 241
PBX, 90, 241
PCM, 136, 241
Permanent Signal, 241
Piggy-Back Twistor, 204, 241
POL: (Problem Oriented Language), 187, 239, 241
Postfix, 193
POTS, 21, 241
PPM: Pulse Position Modulation, 139, 241
Price Umbrella, 218, 241
Primary Center, 91
Progressive Control, 33
PROTEO, 206
PRX, 206
Pushdown Stack, 65, 191, 242
PWM: Pulse Width Modulation, 139, 242

R

Range Extender, 9
Rearrangeable Network, 170, 242
Receiver, 242
Regional Center, 91
Register, 189, 242
Register Progressive (System), 36, 242
Remreed, 65, 242
Reorder, 21, 243

Reverse Polish, 193
Revertive Pulsing, 40, 103, 243
ROTL, 243
RSS, 135, 243
Resistance Design, 9
Rough Traffic, 166
RTA: Remote Trunk Arrangement, 243
RTL: (Resistor Transistor Logic), 65, 243

S

Seize, 243
Selector, 33
Sender, 37, 244
SF: Single frequency, 106, 244
Sidetone, 6, 244
Signaling, 100
Signal Processor, 63, 244
Simulation, 132
Single Frequency (SF), 106, 244
Sleeve, 51, 122
Smooth Traffic, 166, 244
Software, 244
Source Code, 244
SP: Signal Processor, 63
SPC 1A, 184, 244
Station, 5, 245
Step-by-Step: SxS, 32, 245
Strowger Switch, 32, 245
STS Switch, 73, 202, 245
Subroutine, 66, 245
Subset, 245
Subscriber Loop, 9
Supervise, 21, 104, 245
Switch Train, 245
SXS: (Step-by-Step or Strowger Switch), 32, 245
System Century, 203
System X, 204

T

Talking Battery, 6, 245
Talkoff, 108, 245
Tandem (Office), 31, 245
Tariff, 218, 245
TDM, 136
Time-Slot Interchanger, 72, 74, 246
Tip-Ring-Sleeve, 122, 246
TMS, 73
Toll, 31, 68
Toll Center, 91, 246
Toll Point, 91, 246
Traffic, 139, 148, 246
Traffic Density, 246
Traffic Intensity, 140, 246
Traffic Mix, 246
Transit, 246
Translation, 57, 69, 246
Trunk, 4, 246
Trunk Splintering, 64, 247
TSI: Time Slot Interchanger, 72, 74, 247
TSPS, 82, 247
TST Switch, 73, 202, 247
Twistor, 57, 247
Two-Way Trunks, 247
TXE (1, 2, 3), 204

V

Variable Spilling, 69

W

Watch-Dog Timer, 63
Wire Center, 247

X

X-bar: (Crossbar), 41, 247
1A Processor, 63, 248
911 Service, 114, 248